THE COMING DARK AGE

*THE IMPACT OF
THE WAR TO BAN
FOSSIL FUELS*

Steven J Bolen, PRP

Plano, Texas

COPYRIGHT © 2021 BY STEVEN J BOLEN

Notice of Copyright.

All rights reserved. No part of this book may be reproduced in any form by any electronic or mechanical means, including information storage and retrieval systems, without permission in writing from the publisher, except by a reviewer who may quote brief passages in a review.

To find out more, Go to TheComingDarkAge.com

Printed by:
Decisions Consulting

Printed in the United States of America
First Edition, First Printing 2021
ISBN 978-1-63877-281-1
ISBN 979-8-42635-713-6

Library of Congress Control Number: 2021911126

To Shirley,

For the many miles walked while listening to me process all the thoughts that have become part of this book.

Contents

Abbreviations ii
Personal Note 1
Introduction 2
The Coming Dark Age 7
Warning: Confirmation and Other Biases 11
A Glimpse into Life in the Agrarian Age 17
The Green War Plan 20
A Glimpse into Life in the Mechanical Age 47
Human History is Energy History 50
A Glimpse into Life in the Industrial Age 65
The Great Divide 67
Lessons in a Barrel 74
Better Living with Petrochemicals 88
Fossil Fuel Success = Man's Success 120
The True Story of Renewables 139
Can We Electrify Everything? 160
The Big Lesson 181
The Problem with Computer Models 187
The Dark Future 194
The Dangers of the Green Solution 221
Is the Future Dark or Bright? 249
A Proposal 268
Epilogue 273
Endnotes 275
Index 295

Abbreviations

W -> Watt
kW -> kilowatt = 1000 W
MW -> Megawatt = 1000 kW = 1 million W
GW -> Gigawatt = 1000 MW = 1 million kW = 1 billion W

°C	Degrees Celsius
CCS	Carbon capture and storage
CFC	Chlorofluorocarbons
CHP	Combined heat and power
CO_2	Carbon dioxide
COP	UN Climate Change Conference of the Parties
CPI	Climate Policy Institute
CSP	Concentrated solar power
EJ	Exajoule
ERCOT	Electricity Reliability Council of Texas
EU	European Union
EV	Electric vehicle
G20	Group of Twenty
GDP	Gross domestic product
GHG	Greenhouse gas
GND	Green New Deal
Gt	Gigaton
GWh	Gigawatt-hour
GWth	Gigawatt thermal
IEA	International Energy Agency
ICT	Information and communicating technologies
incl.	Including
IPCC	International Panel on Climate Change
IRENA	International Renewable Energy Agency
km	Kilometer
kWh	Kilowatt-hour
LBNL	Lawrence Berkeley National Laboratory
m^2	Square meter
m^3	Meter cubed
MJ	Megajoules
MWh	Megawatt-hour
N/A	Not Applicable

NDCs	Nationally Determined Contributions
OPEC	Organization of the Petroleum Exporting Countries
PJ	Petajoule
PV	Photovoltaic
R&D	Research and Development
RD&D	Research, Development, and Demonstration
SDG	Sustainable Development Goals (UN Goals)
SE4ALL	Sustainable Energy for All
STC	Standard Test Conditions (for solar panels)
T&D	Transmission and distribution
TFEC	Total final energy consumption
TPES	Total primary energy supply
TWh	Terawatt-hour
UN	The United Nations
Unity	Sanders / Biden Unity Plan
USA	United States of America
USD	United States Dollar
VOC	Volatile Organic Compound
VRE	Variable renewable energy
yr	Year

Imperial Prefixes (subset as used in the text):

Unit Multiple	Prefix	Symbol
10^3	Kilo	k
10^6	Million	M
10^9	Billion	B
10^{12}	Trillion	T
10^{15}	Quadrillion	Quad

Metric Prefixes (subset as used in the text):

Unit Multiple	Prefix	Symbol
10^3	Kilo	k
10^6	Mega	M
10^9	Giga	G
10^{12}	Tera	T
10^{15}	Peta	P

The Coming Dark Age
Extra Material Webpage

TheComingDarkAge.com/extras

"The energy industry is the industry that powers every other industry. To the extent energy is affordable, plentiful and reliable, human beings thrive. To the extent energy is unaffordable, scarce or unreliable, human beings suffer." -- Alex Epstein

Personal Note

Have you ever screamed at the television?

Ever sat in front of the television listening to politicians or so-called experts discuss a topic, which happens to be your personal area of expertise? Watching everyone smile and pat each other on the back over the non-sensical answers? Ever feel like they are pushing solutions that don't relate to the problems posed?

This was the position I found myself in during the recent election cycle when the topic turned to the environment, climate change, and proposals to address these issues. I would listen to the discussions and shake my head in disbelief and thus this book was born.

After the second Presidential debate of 2020, I began this project with the intent to set the record straight and to help educate non-engineers on the topic by writing a short book, but all too often in my reading, the engineering alarm bells kept going off in my head with thoughts of "wait, it just doesn't work that way," along with "hey, what about....?" As a result, this ended up longer than I expected, and while there is no end to the material available on this topic, I hope that you will find this to be a reasonable discussion of key elements of the subject and will find yourself educated next time you hear these "experts" talking in the media.

Much of the climate debate is driven by a small vocal minority with specialized knowledge and narrow agendas. I fear that allowing public policy to be driven by such a small minority can lead the nation and world to places it will wish it never went. And the effect of that could be an age of darkness and misery.

Introduction

> *"If man's 50,000 years on this planet are divided into lifetimes of approximately 62 years, then there have been 800 such lifetimes. Of these, over 600 were spent in caves, only the last 70 have had written communication, and only the last 6 have had printed words. But of all of them, the most critical is our lifetime — the 800th. This one lifetime is the center of history with as much happening in it as in all the previous lifetimes put together."*[1]
>
> Alvin Toffler, **FUTURE SHOCK** (1970)

It was a long slow climb of humanity from the ancient past's back-breaking labor to today's technological peak. Since Alvin Toffler wrote in 1970, yet another of his lifetimes has passed. Of Toffler's now 801 lifetimes, just four lifetimes have been spent in the era of extensive coal usage and only two lifetimes in the age of petroleum. Fossil Fuels and all that they have brought to life, both good and bad, have occurred in a flash of human history. Yet, they alone have had more impact on the human population's daily lives than any other discovery.

The 6000 years of recorded history are neatly divided into two distinct categories: the time before discovering fossil fuels and the time of their consumption. In 2013, The Atlantic magazine assembled a panel of experts to create a list of the 50 greatest inventions since the wheel.[2] A review of the expert's list shows that 60% of humanity's greatest innovations have occurred in the time after the introduction of fossil fuels, just the last 3% of human

Introduction

history. As Ronald Stein and Todd Royal state in their book, JUST GREEN ELECTRICITY: HELPING CITIZENS UNDERSTAND A WORLD WITHOUT FOSSIL FUELS, "Nothing has revolutionized human life, eliminated poverty, reduced famine, and taken human choice to unfathomable expectations the way fossil fuels and nuclear have since the beginning of the industrial revolution in the 1850s."[3]

The world of fossil fuels now permeates every aspect of our lives, how we live, how we interact, even our governmental policies. It is so pervasive that it is imperative to everyone to have a reasonable understanding of the issue to participate in the discussion in the public square. If there is one universal fact to remember when discussing the policies behind climate, fossil fuels, and renewable energy, as they say, *"It is complicated!"* Actually, it's really complicated.

The world would have us believe that we sit at a crossroads where we must choose one of two paths for our future. In one direction lies further development of fossil fuel resources, lifting more human beings out of poverty, and improving their lives. While in the other direction lies protection of our environment and a livable planet. Left with such a Sophie's choice, we must choose between leaving some societies behind while the advanced ones enjoy a cleaner world or all societies get lifted to a better life but in a polluted one. A study by the American Petroleum Institute for the 2020 election shows that American voters struggle with this Sophie's choice, insisting on having both simultaneously, with 82% supporting taking all necessary steps for environmental protection while 79% supported developing all forms of energy.[4] Could there be a third path at this crossroads?

Every election cycle in recent years has had environmental issues front and center to any discussion. Time is taken in every Presidential debate to evaluate each candidate's view on the topic and leaving it to the voters to decide which viewpoint would be best.

But frankly, this appears to be "... constant bickering to attract votes from those even less informed than our elected politicians."[5]

Clearly, voters need additional help with energy literacy in order to discuss and evaluate the trade-offs associated with the range of fossil fuel considerations. When various United Nations study groups have created Environmental Education plans and programs, they typically develop a scope of work that defines Environmental Education as something like,

> "... a multitude of processes and activities by which an understanding of the environment is developed and through which caring and committed responses are evolved. It is concerned with knowledge, emotions, feelings, attitudes, and values. It aims to produce informed and responsible citizens capable of playing an active role in all matters concerned with the environment in which we all inhabit."[6]

These organizations are generally concerned with only one side of the equation, the environmental side, and less concerned with the cause and sources of the issue. There is nothing wrong with tackling questions on the environment but focusing only on the environmental side does a significant disservice to those trying to understand the issue's complexities.

Participants in the American democracy face a struggle in finding agreement on which of the paths to pursue. What is the right balance between protecting the environment and providing energy? Do we get the power we need and hope the environment is okay? Or do we protect the environment and hope to get the power we need? Confusion reigns among voters as to the right way forward. Extensive reading of books and articles on the topic of energy can leave one more confused than when they started. When energy policy is being considered, it must be considered in its entirety, including the sources and loads that currently exist, what the future might hold, and how technology will evolve. Any plan to transition to a new energy scheme must accurately cover both the present situation and the possible future ones.

In the research for this work, I have spent many hours listening to the endless arguments and reading all that I could find about energy and climate policy. As I reviewed the various proposals, I realized that many were not considered thoroughly and presented half-heartedly, misleading the audiences. When it comes to the issue of fossil fuels, there is far more than electricity and home heating/cooling to be considered. The debate never seems to reach the most complicated but also most important aspects of the issue.

Gaining energy literacy is critical to the energy debate as Sustainability journal wrote in 2017, it "plays a crucial role because well-informed citizens can support the design and implementation of smart and forward-looking policies. Research has shown that people hold misconceptions about energy, and for young students, these may persist into adulthood."[7] An education focusing on emotions, feelings, and attitudes may fail to help readers gain a complete policy picture. The curious reader needs to get beyond the surface emotions to the deeper facts on the matter. However, in the current environment of information / misinformation, facts / alternate facts, spin / counter-spin, it can be challenging to be certain about the validity of any position on this issue. The curious must not rely upon one or even a few articles to get the whole story. Reading through the current news media reveals that many articles actually say extraordinarily little and merely repeat other articles' words rather than provide valuable facts. It often takes significant digging through the materials and following the footnotes to find the facts behind the pieces.

With the support of the left and the media, there are those that would push for eliminating fossil fuels without considering the impact of such a call. In March 2021, a group of 450 organizations stretching across six continents called on President Biden to end all financing for the fossil fuel industry, with the hope of effectively putting them out of business.[8] This letter states, "There is simply no room left for new investments in long-lived carbon-intensive infrastructure" and that "we need a rapid transition from fossil fuels…to renewable energy by 2030." The groups have signed on to a simple solution, "Scale-up international support for a just transition away from fossil fuels, providing support for workers and communities affected by the transition, for decommissioning and repurposing sites, and replacing fossil fuel with clean energy." As Julia Galef noted that on any topic, "the thirty-second version of an explanation is inevitably simplistic, leaving out important clarification and nuance. There's background context you're missing, words being used in different ways than you're used to, and more."[9]

Is the proposed simple plan enough to fix all the problems?

As Stein and Royal point out, "Humanity was in an abysmal state before widespread use of fossil fuels and the products manufactured from petroleum derivatives; and nuclear-generated electricity when self-government, free speech, and necessities we take for granted did not exist."[10] The era before fossil fuels was a truly dark age in humanity's history. The labor was hard, the winters cold, and the food was scarce. The critical implications of the decisions are such that we must educate ourselves on whether we want to face a coming dark age.

1

The Coming Dark Age

> *"In a dark age, the thread of collective memory and cultural continuity snaps, the ends are lost, and a new thread must be spun from whatever raw materials happen to be on hand."*[11]
>
> John Michael Greer, **DARK AGE AMERICA**

What do you envision for "The Coming Dark Age?"

The term dark age is not one that most would use in their everyday speech. Among non-historians, it typically refers to that era in the late middle ages, mostly in reference to European life between the end of the Roman empire and the rise of the Enlightenment. While historians or academics today would never use the term because of newer discoveries, its use helps us understand the period.

Dark ages throughout history have appeared in various places, largely due to the collapse of powerful empires. When those empires have collapsed, they have left behind a shortage of local cultures upon which to rely. These dark periods often leave little behind in the way of writings, art, or music. The general attitude about the dark ages is that they are "a less-than-heroic time in history supposedly marked by a dearth of culture and arts, a bad economy, worse living conditions and the relative absence of new technology and scientific advances."[12] The reason historians consider a period as dark is a general lack of information about it. Their histories are dark.

Why were the early middle ages dark? The Romans left behind the Colosseum, the Greeks left the Pantheon, and the Egyptians left the Pyramids. But the middle ages residents did little in the way of building and did not leave significant structures behind. The great gothic cathedrals of Europe began construction around the late 1100s to 1200s, at the end of the dark period.

The key to the construction of the aforementioned ancient landmarks was the availability of slave labor. Each was essentially constructed by slaves under the authority of strong governmental states. With the collapse of those empires, the ability to have access to such a slave workforce diminished. In the dark middle ages, the primary occupation was as a tenant farmer on the knight's or the clergy's land. In this era, it took 20 people working the land to feed 22 people. There were just those two people who were freed from working the land, those being the clergy and the royalty. As John Tamny wrote in **WHEN POLITICIANS PANICKED**, "before automation, and yes, global trade, the vast majority of human exertion was directed toward the creation of food. Work was life, albeit not in a happy way. People worked dawn to dusk six days per week in order to survive. Most worked until they died."[13] With nearly everyone dedicated to keeping everyone fed, warm and alive, who was left to be creative, be artists, or invent?

Around the mid-12th Century, there is a warming period (see figure 1), referred to as the "medieval warm period," where the temperatures were a few degrees warmer than they are today. The result of this change was that Greenland could be occupied by Viking farmers, and in general, agriculture was quite successful. Europe emerges from the dark period when the number required to feed those 22 people moves to 17-18 people due to that agricultural success from the improving climate. With the freeing of only a couple of percent of the population from working the land, artist, musicians, and inventors emerge, leaving us with artistic works that fill our museums today.

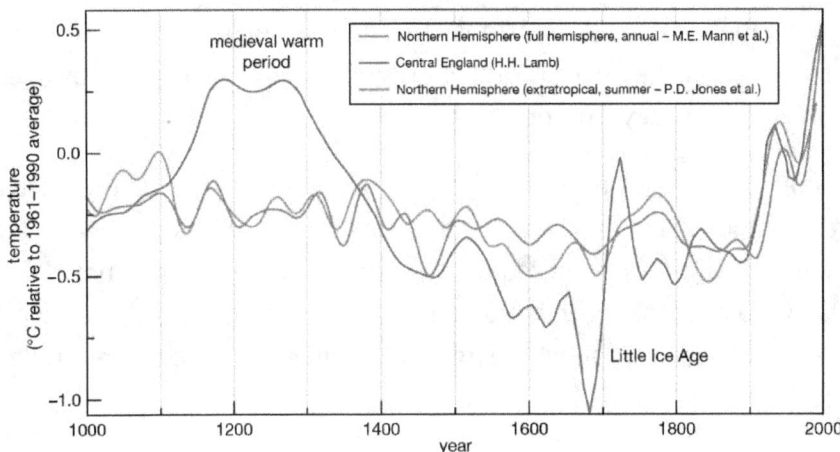

Figure 1 Estimated Temperature Variations for the Northern Hemisphere

In the 21st Century, life has improved as the number to feed those 22 people has fallen to less than 1 person. Twenty-one of those persons have been freed to pursue activities besides just providing for themselves. All around us, the world is full of artists, musicians, and inventors. People engaged in the creative arts now outnumber those who provide for our sustenance. The result is "arguably the greatest gift of economic growth is the growing ability of people to showcase their individual skills and unique intelligence while on the job."[14]

As we will see, the availability of food was the limiting factor for much of man's history, but no longer. Today, the limiting factor is energy. Energy delivered either in the form of electricity or fuels. All those creative people have found ways to make our lives as dependent on energy as on food.

While only 1 person provides food for 22 people, another 2 are now required to provide us with energy. Viewed broadly, 6.4 million Americans are employed in energy generation, distribution, or usage.[15] Of those, 1.9 million work to generate electricity (1.1 million in traditional fossil fuel facilities and 800,000 in renewables). An additional 1.4 million work at energy distribution, from pipelines to gas stations. In our current world, renewables provide about 11% of our total energy consumption but employ over 40% of the energy workers.

Could the energy policies of the future, as a result of increased renewables, change our allocation of workers? We might very well be on the path that will require 20 workers to create the energy for the 22 to use. Taking the wrong approach to energy policy may well result in the situation where energy production is used primarily to produce more energy.

This was the trap in which those of the past found themselves and of which the future risks falling. Fossil fuels have freed us from this dilemma before. From their benefits, life has become easier and longer, safer, and more enjoyable, more freedom, and less drudgery. It is the products around us every day, enabled by fossil fuels which have made the difference between the dark ages of the past and the present.

Warning: Confirmation and Other Biases

> *"Confirmation Bias is the tendency to look for information that supports, rather than rejects, one's preconceptions, typically by interpreting evidence to confirm existing beliefs while rejecting or ignoring any conflicting data."* [16]
>
> American Psychological Association

Before stepping into the subject matter at hand, it is prudent to review the main obstacles to understanding the world's issues: confirmation and status quo biases. These biases are built into our brains and our psychology so that we have the ability to distort what we read and see things in a way that might not reflect reality.

With confirmation bias, "we tend to give more credence to information that supports what we already believe, or that validates decisions we have already made."[17] When faced with social media or research related to a question upon which we have already taken a position, we tend to accept that which agrees with our position and reject all else. The result is that any two people can reach vastly different conclusions when reading the same material because of their pre-existing beliefs. Research has been able to verify this is true by having subjects read articles and asking questions afterward. Still, we have all seen this in action in our current political climate. People from both sides of the political aisle see the same news and take it to support their own views. This bias makes it hard to change minds or to challenge long-held beliefs.

Illustrating this bias is the story of the man who claimed that he was dead. After extensive counseling, his psychiatrist explained at length that dead men do not bleed. Showing him numerous medical texts and taking him to witness autopsies, the man finally caved and agreed that dead men do not bleed. The psychiatrist then took a razor and made a small cut on the finger of the man. When the blood slowly dripped from his hand, he exclaimed, "Good Heavens, dead men do bleed!"[18]

Adding further complication is status quo bias, which is our tendency to prefer a course of not changing over the path of radical change. We experience this bias when the typical family discusses where to dine out and invariably starts with the places they "always" go, rather than any other restaurant selection.

Overcoming these biases takes a concerted effort on the part of the reader. It requires reviewing all the information available by reading entire articles, not just the headlines, and reviewing the sources and the cited authorities' qualifications.[19] Since we live in a world of virtually unlimited sources, especially on the topic of this book, a careful review of the authorities from which you are acquiring your information will help you sort truth from fiction.

Those who are fighting confirmation bias might have noticed the minor issues with the preceding paragraph. Although the information in the paragraph is sound and indeed the way to combat these biases is a careful reading of the documentation, what was the authority? I carefully footnoted my source, an online psychology website which in and of itself was also sound, but if you reviewed the author, you would notice that a freshman Harvard premed major wrote this. The information is not wrong, and her references and sources were sound, but it would have been better to check out her source first (How Confirmation Bias Works)[20].

However, on the downside, as Professor Dan Kahan of the Yale Cultural Cognition Project notes, "For ordinary citizens, the reward for acquiring greater scientific knowledge and more reliable technical reasoning capacities is a greater facility to discover and use – or explain away – evidence relating to their group's positions."[21] We all must be careful not to use our additional knowledge merely to reinforce what we already believe. Further, We must also be careful not to use the newly acquired information to browbeat others who are still struggling to come to terms with the fact that they may have been misled over time. Research has shown that repeatedly exposing someone to the correct information in small bits has a more significant impact than a large, highly technically detailed presentation. Unless, of course, they are an engineer as we tend to love such things.

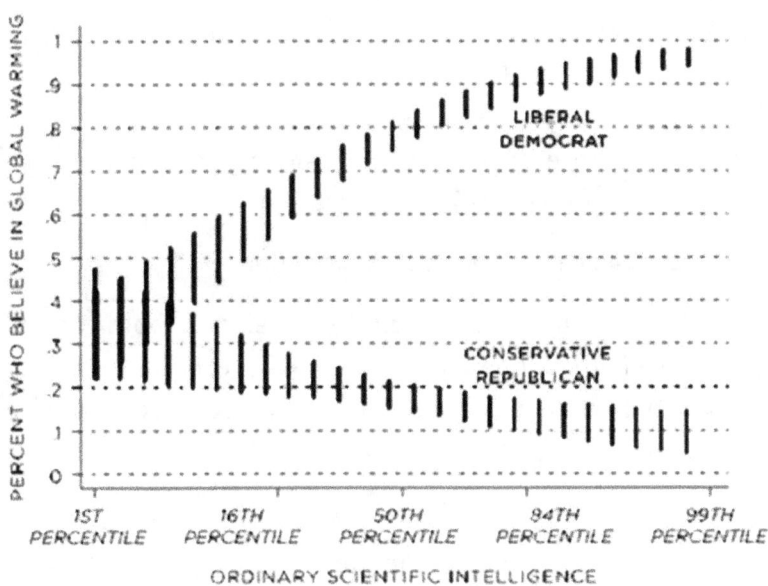

Figure 2 Effect of Science Knowledge on Global Warming Belief

The graph in figure 2 shows that when asked about their belief in human-caused global warming being a threat, how strongly one feels in their position is highly related to the amount of knowledge they have about general science. The more that one knows, the more one is convinced they are correct, and the less one knows, the more they agree with the other side. The more education it seems, the more that everything being read agrees with your position.

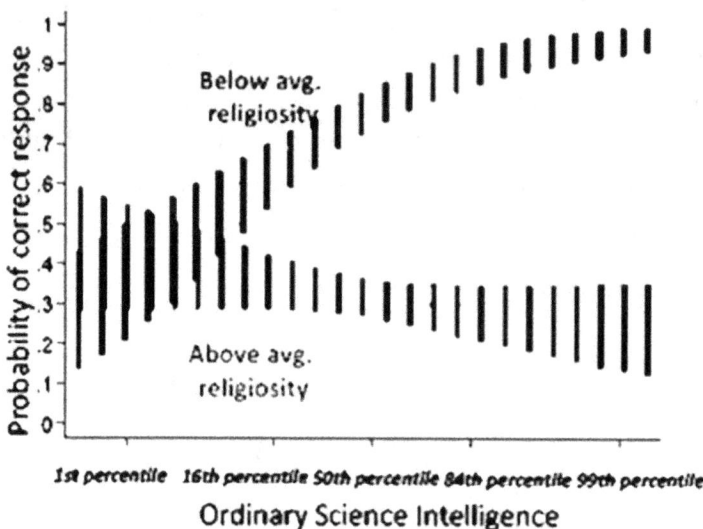

Figure 3 Effect on Religion on Belief in Evolution

Whether it's political belief or belief in a religion, both can affect the response. Figure 3 shows how the same effect occurs when asked about evolution between those with a religious background and those without.

When reviewing sources and authorities, it is critical to dig a bit deeper to understand what is being presented. Consider figure 4[22], it clearly shows the consensus among different government projects as to the warming of the earth. This graph is posted prominently on the front page of the NASA Climate Change website. A website dedicated to proving universal consensus on climate change among scientists. However, the website forces you to dig into the footnotes of subpages to determine that all the sources agree because all four citations are based on the same NASA – Hanson data set.

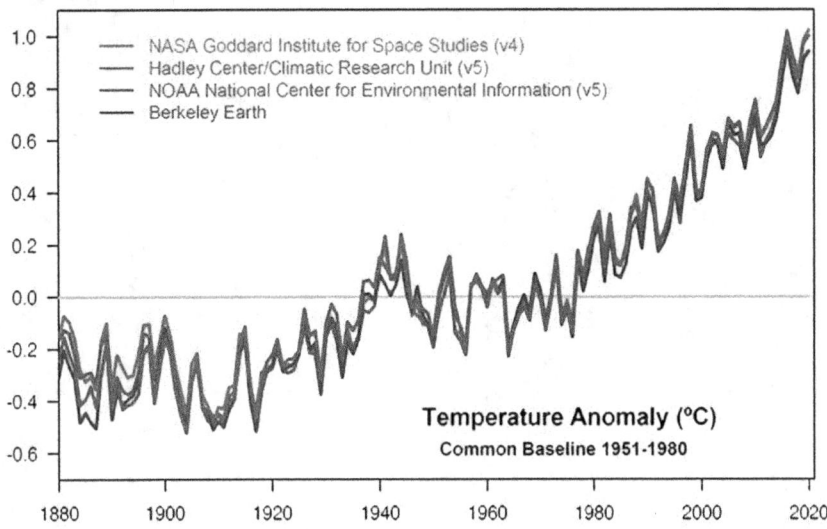

Figure 4 Temperature Baseline

Understanding these graphs can be challenging but important as much of the climate change debate revolves around the many versions and permutations of these graphs. Not understanding or checking the source is rampant. The Army War College used the same chart to claim "A World of Agreement: Temperatures are Rising" when it adds just one new source, Japanese Meteorological Agency, which also uses the same data set.

These are not five different studies but five interpretations of a single data set. Confirmation bias means that there are two responses to this graph, "See, I told you so" and "Yeah, but." But the real question has to become, "so what does this tell us?"

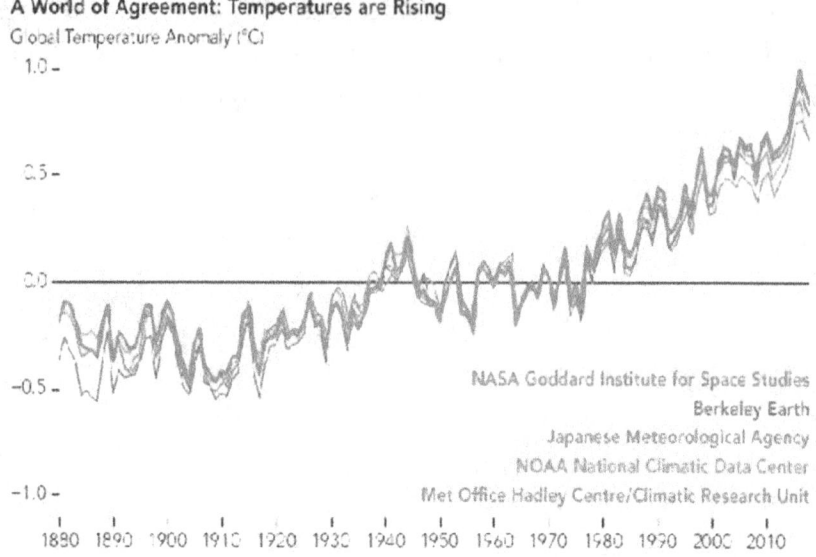

Figure 5 Global agreement according to Army War College

We must approach such data with wisdom and thoughtful consideration, and not just taking the first interpretation that comes to mind. In her book **THE SCOUT MINDSET**, Julia Galef refers to this as motivated reasoning and reminds us, "When we want something to be true, we ask ourselves, "Can I believe this?" searching for an excuse to accept it. When we don't want something to be true, we instead ask ourselves, "Must I believe this?" searching for an excuse to reject it."[23] Having the right mindset and being equally willing to accept or reject information is the key.

I want the readers to know that while this particular topic yields endless rabbit holes of internet material, and thus there is a limit to checking, I at least made a conscientious effort to be accurate. Every effort has been made to use authentic authorities, reliable sources, and unsensationalized information to develop this work.

A Glimpse into Life in the Agrarian Age

Before exploring how humanity should lead future lives, maybe looking into how others lived life in the past might be instructive. It can be challenging for modern-day Americans to picture lives outside of our own experiences, but the future is unlikely to look like today. It will be necessary someday to imagine life differently.

In 1645 during the reign of England's Charles I, two young brothers parted ways, as had so many others throughout time; one, Thomas Snell (1634-1724), headed for the new world with his uncle while the other, Manasses Snell (1628–1714)[24], stayed behind in Fillongley, Warwickshire, England. As landholders of even a small bit of land, the family was better off than the peasants who worked the land as tenant farmers.

Situated between Coventry and Birmingham in Northeast England, the farmland around Fillongley, long cleared of the natural forest, was primarily used to grow various grains and raise sheep and other domesticated animals. While the initial stirrings of the industrial revolution were beginning in nearby Birmingham, the residents of Fillongley lived and worked the land, the same land as their fathers and grandfathers.

As with any agrarian society, whether in Europe or elsewhere, those who worked the land made up over ninety percent of the population. Life could best be described as harsh, uncomfortable, and often relatively short. While both brothers lived exceptionally long lives (90 and 86 years, respectively), the expected lifespans of their tenants were around 55 years.[25] Using only the strength of their backs and the backs of the domesticated animals at their disposal, the peasants worked the land owned and controlled by the landholders.

In order to provide heat and fuel for cooking, the tenants would need to spend hours collecting wood and other materials for building fires. They developed a term for when you gathered your own personal supplies from the natural forest, which was referred to as "winning." Winning wood by your own efforts was in opposition to having to purchase materials collected by others at a significant cost adder. The only affordable way was to win wood on your own. During the long dark nights, the tenants used wood to provide heat within the single room homes they shared with their animals. The earthen-floored homes constructed of sticks, straw, and mud would have a single firepit located in the center of the room so that the walls would not catch fire. At best, the room was smoky, poorly lit, and cold in the wintertime. Crowded around the fire with the farmer would be his wife, who would have upwards of eight pregnancies in her lifetime since 20% of the children would die in childbirth, and upwards of 50% of the peasant children would not survive that first year. Those children who survived to the age of 12 would begin the transition to adulthood, on the path to marriage. Their simple meals consisted mainly of their own cultivation, the grains (rye, oats, and barley), along with bread and vegetables. Rare was a meal that would contain meat, which would only be available when the work animals had perished.

Arriving in Bridgewater, Massachusetts, Thomas acquired land and began farming. With time, in the new world's vast open lands, Thomas expanded his land holdings to include 10 acres of woodland and such open areas that eventually Snell's plain and Snell's meadow became listed on city plots. With time, he was able to add blacksmithing to his skills and opened a blacksmithing shop. His will included the usual stock holdings of a farmer: axe and saws, plough with plough chains, gun, blacksmithing tools, horse and cart, and an "old woman slave" named Maria. He came to the New World with little, but through life as a landowner left a bit more behind for his children, a sum of £1200.

Life in the new world was little different from that back in England, using the fire of the locally collected wood to heat homes while also using the heat for cooking the locally grown meat and produce. Growing what grain and vegetables they could to feed themselves, the farmers would hope for a bit leftover to sell at the local markets.

On both sides of the Atlantic, the land was worked by hand with primitive tools, unchanged in millennia. If a beast was available, it would be used to pull the wooden plows, otherwise it became a duty forced upon man himself. The weeding and tending of the rows of crops meant long hours of work in the fields. The community-wide effort to harvest grain with a scythe was very much the same as had been done for centuries.

Life for the landholder held out hope with the ability to expand their businesses or move on to greener pastures, but for the peasant, there was no hope. Life seemingly would continue as it had for many generations. The energy of life was provided by the grains and vegetables they could grow and the wood they could gather and chop.

But life would soon change for the better. Better tools would soon arrive from Birmingham's metallurgy industry, and with the increase in agricultural productivity, a better world would be on the horizon.

2

The Green War Plan

Decarbonize: (verb) To reduce the carbon released by human activity[26]

> *"Today, global circumstances have changed dramatically. The 7 billion people who inhabit planet earth no longer live in more than one hundred separate boats (countries). Instead, they all live in 193 separate cabins on the same boat. But this boat has a problem."*[27]
>
> Kishore Mahbubani (2014)

Any discussion on the future of fossil fuels must inevitably begin with considering their effect on the climate. Over the years, the name for the climate question has repeatedly changed, from global cooling to global warming to climate change to climate disruption to the climate emergency.

Much debate can be had about how climate change will affect humanity and the earth and whether there is a scientific factual basis for its verification. While much will be discussed here, climate change's fundamental facts are entirely irrelevant to this discussion. What is far more important is the reaction to the belief in climate change within the political sphere and in the world of social media. As climatologist Stephen H Schneider, author of the book **GLOBAL WARMING,** told Discover magazine, "To avert the risk (of potentially disastrous climate change) we need to get some broad-based support, to capture the public imagination. That, of course, means getting loads of media coverage. So, we have to offer up some scary scenarios, make simplified, dramatic statements and little mention of any doubts one might have."[28] It matters little if the effect is real or not; if the social and political decision-making machines have decided it is reality, then the world's authorities will be responding as if it reflected reality, and the subsequent war on climate will drive the coming dark age for society.

As a result of this, the perception of the voting public has changed. Between 1997 and 2007, New York Times / CBS polls found that the number of Americans who believed that "recent weather had been stranger than usual" grew from 5% to 32%.[29] Further, a scientific survey from 2019 by the website Just Facts showed that only 34% of voters believed that the earth had not warmed since the 1980s. Ironically, a dive into the data shows that 83% of the respondents held this belief despite being too young to remember the 1980s. [30] It is enough to believe, even without the experience.

We see this phenomenon in other places in our society. Medical studies show that public use of masks may cause just as many infections as they prevent, but the majority agrees that masks work, so at least through 2020, we wore them. The top 1% of federal taxpayers pay 38.5% of all taxes[31], but the majority believe that they do not pay taxes or not their fair share; therefore, we implement tax policy as if that is true. Other facts like Napoleon was short (he was 5'7," tall for his day), police requiring a waiting period before filing a missing person's report (there generally is not), or humans only use 10% of their brains (every part is busy doing all sorts of things all the time) are often widely believed. Many people would swear these things to be true when questioned and without any curiosity about the truth. The facts on climate change are no different.

It is essential to understand that there is a difference between facts and values. Recently, there has been a trend for people to talk about "their facts" or "their truth" as if they are the same thing. John

Michael Greer, in his book, **DARK AGE AMERICA**, sums this up nicely as "Philosophers generally recognize a crucial distinction between facts and values; there are various ways of distinguishing them, but the one that matters for our present purposes is that facts are collective and values are individual."[32] Values are personal because they relate to how one feels, while facts are those upon which we all observe the same thing. It being a warm spring day is both a fact and a value. There are facts in the statement, such as it being daytime since the sun is in the sky and also, it is spring because of the season marked by the calendar. However, warm is a value as some individuals could consider 75 degrees to be warm while others believe it to be cool. Napoleon was 5 foot 7 inches tall which is the fact but whether he was tall or short largely depends on individual perception. To an Olympic gymnast, Napoleon is tall; to the NBA player, he is short. Confusing facts and values can lead to poor judgment and the making of poor decisions.

Too often, the implementation of government policies reflects what is easiest to explain to the public, and politicians find it easier to communicate values rather than facts. Right now, it is easiest to act as if the environmental crisis is real. Regardless of the actual observed effects, the belief in the existence of the crisis is enough to drive international politics and policy. The discussion of the truth and facts behind climate change is a book-length topic itself, but that discussion is beyond this book's scope. Go to the extras website for this book (thecomingdarkage.com/extras) to access my suggested references on websites and books.

The Green Solution

> *"I would transition away from the oil industry, yes,"* Biden said in the presidential debate's closing minutes under peppering from Trump. *"The oil industry pollutes significantly. ... It has to be replaced by renewable energy over time."*[33]
>
> Second Presidential Debate 2020

Late in the second Presidential debate of the 2020 election cycle against President Trump, President Biden fully committed to a green agenda that would lead to eliminating fossil fuels when he commented that he would transition away from the old oil and gas industry due to their impact on the environment. When questioned, he doubled down on the idea that, indeed, he meant that he planned to shut down the fossil fuel industry as a whole. There was no fully fleshed-out plan, just a plan to "transition away from the oil industry."

Endlessly, debates circle the internet on questions of implementing green policies, most often to no end. These can never end because both sides base so many of their discussions on the logical fallacy of the "strawman argument." The strawman is creating a position for the other side which the other side does not actually hold. U.S. Representative Alan Grayson demonstrated the perfect example of the strawman argument when in 2009, he presented placards with the "Republican Health Plan" marked "don't get sick" and "if you do get sick, die quickly" on the house floor. Clearly, neither side would agree that this was the GOP health plan, yet still, it was presented. Making a reasoned argument for an alternative to strawman plans wastes time and effort if it represents alternatives to something different than the program to be implemented.

There is no single definitive "Green" plan to eliminate the use of fossil fuels, and as such, it is impossible to list what the process will be exactly, but there are several plans in the works. Many programs have been put forward as starting points, and a review of these plans can give some ideas as to what the final Green plan might be. The most complete green strategy is the U.S. House's Climate Leadership and Environmental Action for our Nation (CLEAN) Future Act, whose draft was initially released in 2020. Since the CLEAN Future Act reflects many grants and funding for ideas without actually presenting the details, it is necessary to include some additional specifics. Those specifics are obtained by referring to plans which are more fleshed out proposals, such as The Solutions Project[34], The 2035 Report[35], The Green New Deal (GND)[36], Sanders/Biden Unity Plan (Unity)[37], Greenpeace[38], and the UN Climate Change Conference of the Parties (COP).[39]

Assuming that those items shared among the different plans will ultimately reflect the final Green deal, then the fundamental points of any green plan to eliminate fossil fuels would contain the following items:

1) Decarbonize the Electrical Grid
2) Decarbonize the Transportation Sector
3) Regulate the phase-out of unsustainable sources.
4) Conserve, Recycle and Rebuild buildings and land.
5) Use Banks and Financial Institutions to implement the plan.

In this work, I will use the term **Green Solution** as the identifiable term for the generalized plan, which the green movement would generally agree is the plan as we debate the issues. Since there is no one plan, this is the best we can do.

Decarbonizing the Electrical Grid

When talking about electrical generation, it is essential to understand the terms and differences between capacity and generation. As of 2019, the United States had an electrical generation capacity of 1,100 Gigawatts, of which 24% of the capacity is renewable. However, the actual electricity generated in 2019 was 4.1 trillion kilowatt-hours of electrical power, of which 17% (697 billion kilowatt-hours) came from renewable sources.

Several sources are used to generate electrical power, and many local factors influence which source is used in any given location. The west coast has access to rivers on which to build hydroelectric dams, while Texas has an abundance in other areas like natural gas and wind.

Table 1 Electricity Energy Generation Sources

Source	United States[40]	Texas[41]	California[42]
Natural Gas	38%	47.4%	46.5%
Coal	23%	20.3%	0.15%
Nuclear	20%	10.8%	9.38%
Wind	7.3%	20%	7.23%
Hydroelectric	6.6%	0.2%	13.4%
Solar	1.8%	1.1%	14.0%
Biomass	1.4	0.1%	3.03%
Geothermal	0.4%	Less than 0.1%	5.92%

Renewable sources of electrical power generation have been in use as far back as the construction of the Hoover / Grand Coulee Dams and the Tennessee Valley Authority during the recovery from the Great Depression in the late 1930s. The hydroelectric capacity has remained mainly unchanged since the 1970s, but the addition of wind and solar in the new millennium has increased renewables' total capacity.

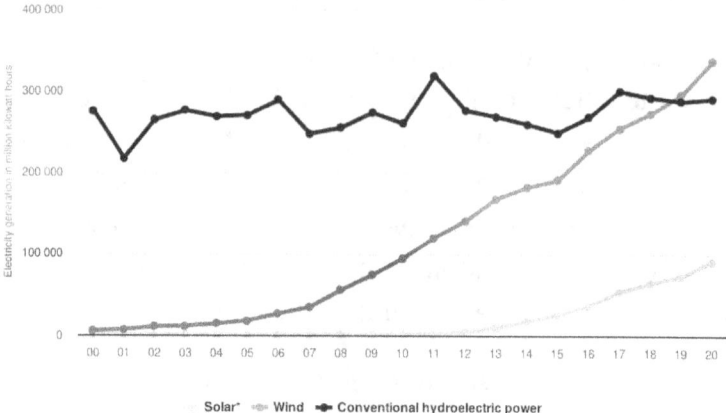

Figure 6 Renewable Energy Sources

As called for by the **Green Solution**, decarbonizing the electrical grid involves removing those sources based on carbon and installing new carbon-free sources. The GND is vague on the numbers but calls for "100 percent sustainable energy for electricity ... by no later than 2030 and decarbonizes the economy by 2050 at the latest fully." The Unity plan has some specifics in its plan detailing to "dramatically expand solar and wind energy deployment through community-based and utility-scale systems. Within five years, we will install 500 million solar panels, including eight million solar roofs and community solar energy systems, and 60,000 made-in-America wind turbines." Notice that the emphasis here is on wind and solar, at the exclusion of all else. As one open letter on the Green New Deal points out, "As the United States shifts away from fossil fuels, we must simultaneously ramp up energy efficiency and transition to clean, renewable energy ... in addition to excluding fossil fuels, any definition of renewable energy must also exclude all combustion-based power generation, nuclear, biomass energy, large-scale hydro and waste-to-energy technologies." The issues with the plan designs mean that these proposals will require that power shortfalls be made up by hydroelectric dams and geothermal power plants despite the objections.

To put the Green New Deal plan into context, the average installed wind turbine currently has a design capacity of 1.67 megawatts (MW).[43] On the typical wind farm, the largest installed turbines can generate 3.4MW, with the largest currently available turbine generating 10MW. There are currently around 69,000 wind turbines in operation, providing just under 122 Gigawatts of capacity.[44] So the additional 60,000 turbines would provide an additional 105 GW of capacity and, when running at the usual 33% capacity factor, will generate 33 Gigawatts[1] (GW) of electricity. Likewise, the 500 million solar panels will produce about another 25 GW[2] of power. The combined installed available power would be around 59 GW from this plan. Nowhere close to what would be required.

The 2035 Report Plan aims for 90% renewables, while the Solutions Project aims for 100% renewables. The Solutions Project calls for 496,000 additional wind turbines, assuming that those units' rated capacity will increase from the current average of 1.7 MW to a new generation of 5 MW turbines. These units would provide an additional 2485 Gigawatts of power capacity, enough to replace the existing grid with the expanded use of electricity.

The U.S. Government website, windexchange.energy.gov, indicates the total potential wind capacity could be as high as 10,640 GW (1,300 GW of which would be in Texas), but that would require the placement of 2.2 million 5 megawatt turbines, at a density of 3 megawatts per square kilometer.[45] It was completely unrealistic, but someone spent the time calculating and creating an interactive map of the idea on the internet.

Two items that must go along with any conversion to renewable energy are storage methods for when solar and wind are not generating and an intelligent grid to handle the loading changes

---

[1] 60000 turbines * 1.67MW *0.33 capacity = 33, 066 MW
[2] 400-watt panel at 6 hours of sun with a capacity factor 50%

that an intermittent power source and storage system would create. Interestingly, the **Green Solution** has little to say about these two essential elements.

Ultimately, after the shutdown of fossil fuel facilities, the **Green Solution** grid would consist of wind, solar and hydroelectric generation with some form of energy backup, either in the form of natural gas plants or electrical storage. Many factors determine whether this is possible, as so many of these different proposals are interacting pieces of a puzzle. Can we decarbonize the grid? More needs to be understood before we can answer that question. (Far more on decarbonizing the grid in Chapter 9: Can we Electrify Everything?)

Re-envisioning the Grid

There is in the Green New Deal and in Bernie Sander's version, a line that is missed by most readers of the proposal, "We will spend $526 billion on a modern, high-volt, underground, renewable, **direct current**, smart, electric transmission and distribution grid that will ensure our transition to 100 percent sustainable energy is safe and smooth. (emphasis added)"[46]

The late 1880s saw the introduction of the "War of the Currents" between Nikola Tesla and George Westinghouse on the side of Alternating Current (AC) and Thomas Edison on the side of Direct Current (DC). Each side proposed a differing concept on how to construct the initial electrical grid. The AC approach was favored by those offering a network built around large remote power generation facilities, while the DC approach was favored by those expecting to develop locally generated power as its transmission is limited to a distance of a couple of miles. In the end, Tesla and Westinghouse won the day, and the result is that 140 years later, we have AC power transmitted across long power lines and into our houses.

The earliest applications for electrical power, washers, dryers, furnaces, and light, all ran directly on alternating current power, so having an AC-powered network made a lot of sense. However, today those applications are controlled by digital circuits and, as such, no longer run on AC. Still, like any device without a motor such as phones, televisions, and clocks, they run off of DC power, so each device must contain a convertor circuit that converts the AC to the required DC power. Having a DC supply to the home makes a lot of sense, especially when considering that solar generates direct current power. While wind turbines do create AC power, it isn't at the frequency nor in sync with the grid; therefore, it must be converted to DC and back to AC before being fed to the grid.

The green vision for the power grid would divide the grid into two parts, an AC portion to provide the power across long stretches of land from the remote power stations and a DC portion to provide power to homes through a series of small neighborhood-sized grids. Suppose direct current was provided into the home. In that case, significant power savings could be realized from not needing the large number of AC to DC converter circuits currently present in every device or wall charging block.

With their intended build of more than half a million wind turbines, the Solution Project is further proposing a national wide grid of DC power so that wherever the wind is blowing can provide power to those parts of the nation where the wind isn't blowing. This would be a secondary national grid of power lines from the current AC power lines to be phased in over the next 35 years.

Much has been made of the proposed "weatherization" and energy efficiency improvements that the **Green Solution** suggests implementing in every house and building in the country (see GND). However, in light of the proposed change in the grid, this makes complete sense. Every building with electrical connections will have to be modified to handle the new DC grid, with all the appliances and installed equipment base needing to be updated with a different design of wiring and plugs. Changing the common electrical usage from AC to DC is a major undertaking, not just the plug-n-play that modern technology has taught us to expect.

Decarbonizing Transportation

The transportation sector of the economy involves many elements, from personal to commercial. Fossil fuels are the basis for most methods of transportation in the modern world. The world supports 1.2 billion road-bound vehicles, from individual cars to light and heavy trucks, which move all the goods to the world's markets.

Fossil Fuels also support:

- 39,000 aircraft consuming 225 million gallons of aviation fuels per day to move 4 billion passengers per year.
- 300 cruise liners consuming 80,000 gallons of fuel per day to accommodate 25 million passengers per year.
- 60,000 merchant ships consuming 200 tons of fuel per day.

Decarbonizing means moving away from gas-powered vehicles to a fleet of electric vehicles on the consumer level. On his first day in office, President Biden started the process by signing an executive order to move the federal vehicle fleet to be all-electric by 2024. The move to a 100% electric fleet involves creating a nationwide network of charging systems that can charge fast enough to satisfy customers (minutes rather than hours). As provided for in his infrastructure proposals, the plan would be as transportation is decarbonized, then the network of pipelines, refining, delivery, and the 168,000 retail gas stations would be replaced by retail stations providing electrical charging. When vehicle fuel is no longer available at the retail centers, the use of those vehicles will cease.

The public and commercial side of the decarbonization equation presents significant obstacles. Two different programs for converting public transportation will be required, one for local service and one for long-distance service. The local program will require the modifications of buses to either self-contained or wire-bound electrical systems. A push to subways and regional electrical light rail are reasonable solutions to this problem. In the case of long-distance transportation, the current state of the art in electrical power means replacing privately-owned aviation with publicly owned high-speed rail, at least on a regional level.

On the commercial side, the decarbonizing of long-distance trucking and shipping will be an obstacle that will have to be handled by newer technologies or economic changes that reduce these services' usage. The CLEAN act provides grants and funding for research projects to find an answer to this problem.

Regulating the Unsustainable

While new technology carbon-free sources are being brought online to the grid, older carbon-based and non-sustainable sources (coal, gas, nuclear) will be phased out and decommissioned.

Major international anti-nuclear groups have pushed for the shutdown of nuclear facilities since the accidents at Chernobyl and Fukushima due to safety concerns. The **solution** includes the decommissioning of the 20% of America's total electrical generation capacity created by 96 nuclear reactors located over 58 facilities in 29 states.[47] Internationally, Belgium, Germany, Spain, and Switzerland plan to decommission their reactors before 2030. There are currently 17 reactors already in the process of decommissioning. Due to the nature of the nuclear reactors, they tend not to be affected by external forces such as weather or fuel supplies. The units have been the most consistent source of electrical power in the United States, generating their consistent

20% of electricity since 1987. Around the world, 32 countries operate a total of 440 reactors giving the world a capacity of 400 Gigawatts[48] out of the total 7250 Gigawatts of capacity worldwide.[49] While nuclear plants represent 5% of the capacity, the reactors provide 10% of the generated power.[50] Currently, outside the United States, some 53 nuclear reactors are under construction with a potential capacity of 54.7 GW.[51] Nuclear electricity generation plants have become effectively impossible to build due to regulations, with the last nuclear plant coming online in 1996.

Within the United States, coal-fired power plants have been increasingly less cost-effective than natural gas and other sources, which has resulted in the announcement of the shutdown of 13 such plants in 2020.[52] Through 2020, 61 Gigawatts of generation capacity have been shut down, with an additional 18 Gigawatts of capacity already announced and in the planning stage. Programs like the Sierra Club's Beyond Coal aim to shut down the remaining 191 American coal-fired plants by 2035.

Beyond just shutting down non-green electrical generation, as President Biden indicated in the debate, the **solution** is designed to shut down the fossil fuel industry altogether. Organizations like the Washington DC based Greenpeace plan to use regulations as a means to accomplish this task by implementing four key steps: Stopping U.S. Oil Exports, Stop leasing on federal land for exploration, end fossil fuel subsidies, and end permits for any new infrastructure build.

Energy Requirement Reduction

Every realistic review of the decarbonized economy reveals that there will be a need to use less energy generally. The 2035 Report and Solutions Project each specify the need to reduce energy use by 25 and 40%, respectively. Specific plans proposed to reduce energy usage by providing funding and upgraded building codes to weatherize homes, offices, and buildings and expand recycling programs.

The GND calls for significant funding for grants to weatherize and upgrade buildings such that Americans "fully end[s] all fossil fuel use in buildings by 2030."[53] It is believed that "deep weatherization retrofits will reduce residential energy consumption by 30 percent," which is partly achieved by replacing inefficient forms of housing such as mobile homes with more efficient zero-energy modular homes. Since single-family homes are exposed to the weather on all sides, the use of multifamily housing arrangements where the exposure to the elements is reduced will also improve that energy efficiency. Where possible larger multifamily housing blocks are encouraged. The plan to remove fossil fuels from homes is accomplished by going fully electric in the homes and removing all use of oil, propane, and natural gas for heating, cooling, and cooking. These efforts cover all 108 million single-family homes, 43 million apartments, and 6 million office buildings.

According to the American Gas Association, 62 million (56%) of American homes use natural gas for heat, representing about 23% of all-natural gas consumed nationwide.[54] There are nearly 500,000 miles of pipelines that deliver natural gas and petroleum products to the processing facilities and an additional 2 million miles of pipelines to deliver natural gas to individual customers.[55] In addition to natural gas, 5% of homes are heated by propane and 6% by fuel oil.

Both small family and corporate farming consume large amounts of fossil fuels while generating significant greenhouse gases. According to the U.S. Department of Agriculture, as much as 15% of all agricultural production costs relate to energy usage. The needed investments in decarbonized farm machinery and improvements in building construction and heating will keep farms in line with the requirements of the **solution**. Therefore, a push to invest in ecological farming through conservation and sustainable agriculture will be required.

A New Green Revolution

The term "Green Revolution" was coined in the 1960s by William Gaud, Director of the United States Agency for International Development (USAID), to refer to the work of Dr. Norman Borlaug, who since the 1940s had developed plant varieties and techniques that led to the feeding of an ever-expanding world population. This revolution was "green" in the sense of green plants growing in places, mainly underdeveloped and developing nations, where plants have never grown well, not in reference to ecology or the political movement.

The book market of the 1970s was flooded with books like **POPULATION BOMB, FAMINE 1975!, COUNTDOWN: OUR LAST, BEST HOPE FOR A FUTURE ON EARTH?** and **TEN BILLION**, which told the world that it was not going to be possible to feed all the masses of population that was expected between then and now. However, Dr. Borlaug's work, among others, paved the way for massive improvements in crop yields across the world, and when teamed with GMO products, the world has succeeded in feeding 7.8 billion people with food leftover. Today, the world still has hungry people, not because there is no food for them, but because it is difficult to get them food due to local conflicts and the lack of infrastructure. The Green Revolution used improved crop varieties and improved irrigation and fertilization techniques to increase crop yield from 3 to 5 times what they had traditionally run. When Borlaug was awarded the Nobel Peace Prize in 1970, he was recognized as "the man who saved a billion people."

Figure 7 The Effect of the Green Revolution on Cereal Production

Now the IPCC has a new "green revolution" in mind for the planet. The IPCC Working Group III co-chair, Jim Skea, notes that land plays a vital role in the climate system. Agriculture, forestry, and other types of land use account for 28% of human greenhouse gas emissions. At the same time, natural land processes absorb the carbon dioxide equivalent to almost a third of the carbon dioxide emissions from fossil fuels and industry."[56]

When it comes to actual land usage, it turns out the land taken up by urban areas amounts to about 3% of the total landmass, half of which are roads and parking spaces.[57] Of the remaining 97% of available land, 17% is essentially unusable as rocks, deserts, salt flats, and the like, with an additional 10% of the land being ice and glaciers, that leaves 68% of the globe's land inhabitable and usable for agriculture, forest, and preserved lands. Agriculture from cropland to pasture uses 50% of the remaining habitable land, with Forest (37%) and Scrub and Freshwater(13%) being the remaining parts. Irrigation of crops and the watering of livestock are an enormous drain on natural resources, including 70% of all the freshwater.

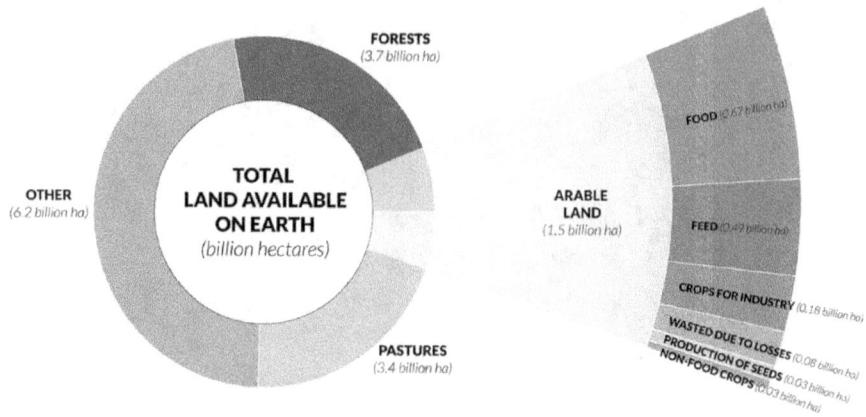

Figure 8 Global Land use for Food Production

In 1993, the Convention on Biological Diversity, signed by 168 countries, committed to setting aside 17% of ice-free land worldwide for nature preservation and protection. To date, 14.7% has been allocated by the signers to the convention.[58] In February 2021, a group of conservation groups, including BirdLife International, Conservation International, The National Geographic Society, the Natural Resource Defense Council, the Nature Conservatory, and others have joined President Biden's America the Beautiful Plan and the IPCC's recent report, "Making Peace with Nature," in calling for amending the Convention of Biological Diversity goal to 30% of land preserved by 2030 (about twice the size of Texas) and 50% by 2050.

Ironically, in their letter, the conservation groups express concerns about man's encroachment into the preserved lands. That the parks are under threat of becoming "paper parks" where their protected status is routinely ignored. It is ironic because the same groups call for increased protection areas, which will by default threaten and infringe on the land currently used for agriculture. The IPCC's report recommends reducing the land usage for cropland by 20% and by livestock by 50%. Further, the report recommends reductions in the use of fossil-fuel fertilizers, pesticides, herbicides, and outright bans on GMO products, while indicating the need to produce additional crops for fuel, fibers, and feedstock chemicals.

All of this will eventually reduce the amount of food being produced. The **Green Solution** to this issue is a push toward vegetarianism and tight control over calorie consumption. A recent USDA report, "The Role of Fossil Fuels in the U.S. Food System and the American Diet," proposes possible future food diets, including the Realistic Healthy Diet (what they think the population might actually be willing to try) and the Energy Efficient Diet (if the scientist had their way). These diets cut consumable calories nearly in half for the average American. Clearly, western diets can never be made available for underdeveloped or developing nations.

As the Guardian points out, livestock provides roughly 18% of our calories yet takes up 83% of farmland.[59] Reducing the consumption of meat, as pushed by Bill Gates, among others, will free up the pastureland for preservation. The Biden climate plan introduced in April 2021 calls for a 50% reduction in meat consumption and a 90% reduction in beef. The end result is that each American would get 4 pounds of beef per year, roughly one quarter-pounder per month.[60] According to the USDA, of the typical 2,500 calories in the American diet, 17% of those calories come from meat, about 440 calories per day. An additional 16%, 420 calories come from a second source, also under attack in the interest of our health, added sugars and sweeteners. Overall, an additional 860 calories will need to be made up from grains and vegetables at the expense of their land demand.

Green Finances

On May 16, 2021, the International Energy Agency dropped a bombshell when it suggested that "there should be no new oil and gas investments after 2021."[61] When the world's nations agreed to the Paris Climate Accord, they also agreed to significant changes to how the world handles financing regarding energy production. In

January 2020, Larry Fink, CEO of Blackrock Investments, the world's largest asset manager, announced that Blackrock would divest from any coal-related business in all but a few specific funds. They observed that "As we move to a low-carbon world," that they would not be a "passive observer in the low-carbon transition."[62] Not to be outdone, Citibank soon afterward announced a new strategy related to ESG (Environmental, Social, and Governance) standards to measure and manage investments based on scoring concerning the low-carbon transition. These two banks are just two of a group that represents the majority of banks, which have signed on to the Partnership for Carbon Accounting Financials (PCAF), a set of accounting standards that allow a financial company to evaluate any proposed business based on ESG scoring. Over 85 financial institutions with $17 trillion in financial assets under control have signed on to the standard. The PCAF standards cover six major asset classes, including Listed Equity and Corporate Bonds, Business Loans, Project Finance, Commercial Real Estate, Mortgages, and Motor Vehicle Loans. While the first four assets are generally related to business and higher-end investors, the final two items affect everyone who wants to buy a home or a car.[63]

The PCAF standard (see Figure 9) is designed to look at many aspects of an individual or business in order to determine the score. Important issues include the amount of electricity and heat/cooling elements for personal use, purchased goods and services, amount and type of transportation, waste generation, and owned investments.

In April of 2021, President Biden and special climate envoy John Kerry announced that they would be releasing an executive order to have the Security and Exchange Commission (SEC) prepare rules requiring that all publicly traded companies report their emissions and climate-related risks.[64] Already, Merrill Lynch has been giving their investment customers reports about the emissions profile of the companies in which they invest. This is a subtle nudge for the investors to consider moving away from high carbon companies.

For example, how will the scoring for electricity usage be done? The average American household uses 10.7 Megawatt-hours[65] annually. In Europe, that number is 8.4 Megawatt-hours. At the same time, globally, the average is roughly 3.5 Megawatt-hours, although that is just for the 87% of global households with electrical service. Such statistics may well become the benchmarks for whether future loans are available. Why shouldn't American's be expected to reduce our energy usage to match that of the Europeans or the global standard?

Clearly, driving an electric vehicle or better yet, using mass transportation while staying away from purchasing goods that negatively impact the environment will improve your odds of getting a mortgage and likely reduce the rate you will pay.

The CLEAN Future Act provides for the funding of a National Climate Bank to provide "financing for low- and zero-emissions energy technologies" in the areas of "climate resiliency; building efficiency and electrification; industrial decarbonization; grid modernization; agriculture projects and clean transportation. The national bank's goal is to facilitate regional and local green banks' opening to direct funding.

The idea is that if the fossil fuel industry, or as President Biden prefers to call it, "the oil and gas industry," could be starved of financing and the ability to access even basic banking services, then the political goals could be accomplished without having to actually enter the political arena itself. While the public may picture the so-called *Big Oil* exploring for oil, the drillers are actually 9,000 independent oil and gas producers with an average of just 12 full-time employees and just $8 million in annual capital expenses. A survey of these independents by the Independent Petroleum Association of America indicated that slightly more than 50% of companies "accessed external capital markets where more than half of their capital came from private equity or private debt."[66] While trying to starve Big Oil of funding might prove difficult, starving the small independents would be relatively easy, especially with the PCAF standards in place. Punish their sources of private equity or debt for providing financing, and the independents won't be able to drill. And if the independents cannot drill, America is starved of the 91% of American wells which they develop and the 83% of America's oil which they produce.

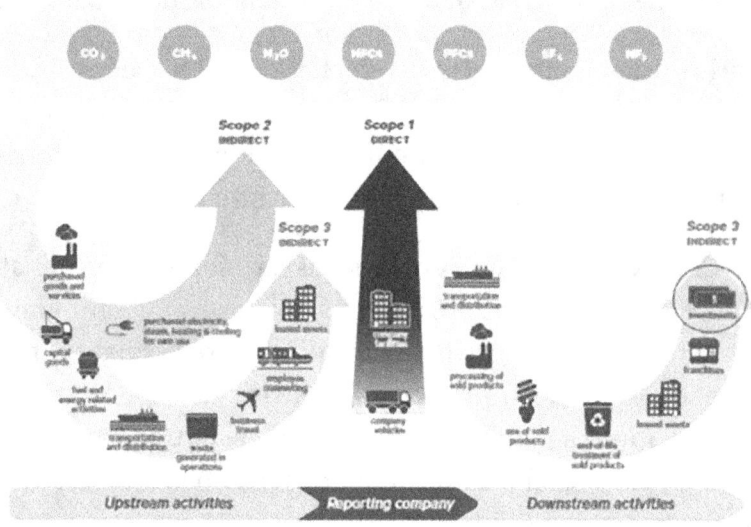

Figure 9 PCAF Evaluation Considerations

Zero or "Net" Zero?

Often, scientific or technical discussions are riddled with field-specific jargon that expresses a clear meaning to those in the know but can be extremely confusing to the casual reader. While the proposal may be clear in its intention to those who make it, the typical reader may not understand what the proposal is intended to accomplish. The plans to eliminate carbon are no different when the terms "zero carbon" and "net zero carbon" are used. Proposals may look very similar and use these terms in ways that they look the same, but there are significant differences. Does the plan eliminate carbon or merely offset the carbon emissions with actions somewhere else by someone else? Many of the recently announced virtue signaling by corporations reveal promises to achieve net zero carbon emissions because they believe that they can pay for someone else's project to absorb carbon and offset their emission, for a price, of course.

ExxonMobil, the largest oil and gas company, has expressed support for the 2015 Paris Climate Accord and promises to achieve net-zero carbon by 2025.[67] This will be done by combining emissions reduction by eliminating flaring and the capture of escaping methane, with investments in projects to reduce the use of energy by consumers. The company itself never gets to zero, but they count the emission savings that customers will ultimately make through the projects they help fund.

In 2015, Bill Gates joined with other venture capitalists to create an investment fund called Breakthrough Energy to help fund innovations and projects with the intent of "Working to achieve net-zero emissions,"[68] according to their website. However, this website adds to the confusion as it also mixes the concept of zero emissions with that of net-zero.[69]

The classic action taken so often to offset emissions is to pay to have trees planted in a third-world country. It is hard to know what they believe happens to those plants after they are planted. They generally are not paying for the land the trees occupy or for maintenance of the tree farm. A 2016 German study indicated that only 2% of the offset projects had a high likelihood of having an environmental impact, while 85% failed to produce any carbon capture.[70]

The band Coldplay paid to have 10,000 mango trees planted in India to offset the emission from its second album release.[71] Of those plants, few survived, and the money for water and support, while promised, never arrived. In another case, the Vatican paid a Hungarian company KlimaFa, to have a million trees planted for which they received the "certificates," but no such trees were ever planted.[72] In many other cases, the trees were planted and grown for about five years before they were cut down and used as firewood by the local residents, and new trees were planted in their place, paid for by another carbon offset project. Other Net-zero programs are even more questionable, such as offsetting carbon by funding programs for non-profit organizations to provide education on climate change issues, a method publicly used and endorsed by both Bill Gates[73] and Bono.

Philanthrocapitalism is the name for the business model which Gates and Bono, and others are undertaking. Philanthropic gifts help start programs while simultaneously they invest in the companies that run those programs, and as such, they intend to make significant monies in the long term. Net-zero games are much more profitable than the zero-emission goal. With net-zero, there will always be emissions to offset, and ever more money to be made and scams to execute.

Amazon proclaims their net-zero policy on numerous television ads, announcing they will be net-zero by 2040.[74] Curiously, an investigation of their website on their plan reveals that the project has such confusing statements as, "Making all Amazon shipments net-zero carbon through Shipment Zero, with 50% of all shipments net zero carbon by 2030." Significant funds are being spent purchasing and advertising their purchase of electric vehicles, even though those vehicles are limited to the last few miles of the delivery process. They further get to net-zero by "Investing $2 billion to support the development of technologies and services that reduce carbon emissions and help preserve the natural world." Like many companies, they are buying their way to the claim of being environmentally friendly. It must be telling that Amazon notes that they signed on to The Climate Pledge, which requires that their offsets will be:

> "*Credible Offsets*: Signatories must take actions to neutralize any remaining emissions with additional, quantifiable, real, permanent, and socially-beneficial offsets to achieve net zero annual carbon emissions by 2040."

A further trick in how net-zero plans will accomplish that feat is that some projects allow for emissions now and plan for negative emissions at a point in the future based on technologies that do not exist, but they expect to exist in the 2040s or 2050s.[75] As Professor Duncan McLaren states, "Net-zero plans that rely on promises of future carbon removal – instead of reducing emissions now – are, therefore, placing a risky bet" and therefore, "determining how safe it is to bet on negative emissions in the second half of this century to avoid dangerous climate change should be among our top priorities."[76]

For simple clarity, a zero-emission project will place no carbon into the atmosphere. In contrast, a net-zero project may put carbon into the air, as long as it promises to remove that carbon somewhere else or sometime later.

Energy Inequality

Consideration of environmental action is essentially a consideration between the Haves and the Have-Nots. The United Nations divides the world's nations into the more developed, the developing, and the least developed countries. The question is how the future will appear. Will the least developed countries be more like the developed ones, or will the developed ones be more like the least developed. Or will the status quo be maintained, along with its energy inequality?

The most developed countries (MDC) used fossil fuels and all the advantages of the energy to rise to the top of the heap. However, if the world hopes to keep CO_2 emissions under IPCC standards, the lesser developed nations cannot follow the same development path. There is no provision within the **Green Solution** for these countries to emit additional CO_2. The proposed idea under the IPCC is to encourage these underdeveloped nations to move directly to renewable energy without the fossil fuel stage of development. This development method is called "leapfrogging," and the plan is to move directly from historical usage of wood and natural sources of energy to renewables like solar and wind.

Most of the world has experienced the concept of leapfrogging when it came to telephone connections. While in the United States, usage of landline telephones peaked at over 90% of all households before giving way to the cell phone market, this was unique among nations. In Africa (see Figure 10), the landline market was never significant outside of industry and government. The general population skipped the landline phase that America went through and went directly to cellphone-based phone service.

While some activists like Sir David Attenborough advocate maintaining the current balance of energy usage, others have a more egalitarian approach. Julia Steinberger, Professor of Social Ecology and Ecological Economics, said, "There needs to be serious consideration to how to change the vastly unequal distribution of global energy consumption to cope with the dilemma of providing a decent life for everyone while protecting climate and ecosystems."[77]

Proposals like the **Green Solution** are statements made by the Have's, in which the Have-Not's reap either the benefits or the consequences. Blanket bans on fossil fuels or certain energy types will indirectly affect those Have-nots when either the fuels, the machinery to utilize those fuels, or the compatible technologies are unavailable. For instance, if the developed countries move to direct current grids, while the developing world still relied upon alternating current ones, those developing nations would soon find their grid incompatible with the latest technologies and suffer from a lack of devices to plug into their grid.

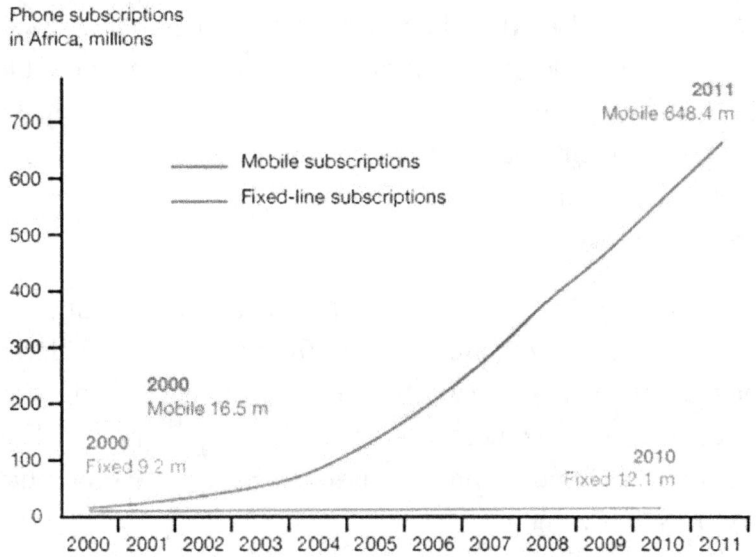

Figure 10 Telephone Subscriptions in Africa

The **Green Solution** presented here represents an honest, reasonable expression of those plans under which President Biden promised to "phase out fossil fuels." Every effort was made to ensure that no element of the discussion represented something proposed by a fringe group or did not represent the mainstream of those who lobby and agitate for a new solution.

It would be wrong to present a version of the **Green Solution** that was merely national or international in scope and not reflective of the local boots on the ground activists. Rest assured that this is not the case. In my hometown paper, just this week, was a front-page article on demonstrations by the Texas Campaign for the Environment (TCE) aimed at Liberty Mutual, stating, "Big insurance companies like Liberty Mutual are basically essential for making sure that fossil fuels projects go forward.... We'd like to see them, one focus on keeping all the stuff in the ground and then two, invest in and start focusing on renewable energy and other climate projects, building renovations and things like that."[78] In response, Liberty Mutual promised that they were, "taking action and have a long-term strategy of decarbonization and investment in renewable energy." Even while corporations work on divesting from fossil fuels, the movement has spread. Several local city council candidates have promised to divest their city's funds from any such offensive investments.

Likely, there are many more ideas and proposals, but many of them are too far outside of the mainstream of thought to be considered or are beyond practical consideration. Furthermore, the proposals presented here are of the most interest as they represent the ideas for which significant unintended consequences are the topic that this book aims to highlight.

The **Green Solution** was always intended to be a war on the fossil fuel industry, hoping for a different future. But as Francis Bacon once said, "Hope is a good breakfast, but a bad supper." Is the hope of the **Green Solution** enough to bring a bright and positive future for all mankind? Or will that hope run out in the long run?

A Glimpse into Life in the Mechanical Age

The age of mechanics is the stage of development when life goes beyond the sheer muscular power of man and beast and begins to take advantage of the knowledge of physics to improve the everyday life of the common man. Using new developments in the metallurgical arts, men soon learned to construct all sorts of new tools from surveying equipment, clocks, and various devices of mechanical advantage. It became the age of the engineer as we know the term today.

While the stirrings of an industrial revolution were happening in the urban centers, for the 85% of the population who remained in agricultural work, life was in many ways essentially unchanged from the agrarian age. But better prospects were ahead.

It is estimated that in 1830, some 250 to 300 hours of labor were required to produce 100 bushels of wheat with the tools at hand, those being walking plow, brush harrow, sickle, and flail.[79] Improvements like the 1819 introduction of the iron plow with interchangeable parts helped. But the real breakthrough came in 1837 when John Deere created his original steel plow, "The plow that broke the plains." A plow with new materials and a new design that, while still pulled by animal teams, made plowing the unbroken ground of the plains possible. With the addition of the McCormick mechanical reaper and numerous threshing machines, by 1860, those same 100 bushels of wheat could be produced with about 150 hours of manual labor.

In the early 1840s, William Bolin and his two brothers Samuel and Benjamin left Virginia by means of a horse-drawn wagon for the west. For their service in the War of 1812, the brothers were granted land claims in central Ohio. Once settled, the brothers founded a new town named for the grist mill they built, Bolin's Mill, located near modern-day Athens, Ohio.

On their plot along the Raccoon River, the three brothers constructed their mill, using a water wheel to drive belts that would provide the necessary power to grind the grain harvested from the surrounding fields. The next generation of Bolins added a lumber mill to the complex by extending the belt-driven system to include powering of sawblades. The power of the water from the river kept the granary and lumber yard in business for four generations, and the town didn't fade into obscurity until well into the 1900s.[80]

While Bolin's grist mill used waterpower to drive belt mechanisms to move millstones, textile mills used the same technique to power the spinning and weaving machinery. The mills used mechanics to improve productivity and eliminate human labor as the limiting factor in production. Industrial development, often limited to access to water resources to drive the mechanisms, continue to grow even if constrained in their placement.

While biological materials continued to be the dominant energy source in the form of nutrition for man and beasts, along with wood and hay for heating and cooking, the mechanics of the era helped reduced the back-breaking labor. Increasingly, the use of driven mechanics helps produce a variety of tradable goods so that it was no longer necessary for every man to be fully self-supporting. While trade had always existed, with mechanics, it can become the main aspect of the economy.

The cities' residents started to acquired conveniences that looked more modern, from horse-drawn buses and trolleys to city gas for lighting. Such advancements were not available to a majority who were still rural residents.

Figure 11 Textile Mill

Life in the mechanical age improved over previous times as the improvements in diet and reduction in labor effort extended life spans for field hands by 5 to 7 years while the average height gained two inches and deaths in childbirth decreased. The long-term trend towards smaller families began as the children were more likely to spend their youth being educated than being used as farmhands.

While much of the Bolin family stayed in Ohio, spreading across the neighboring counties, my 3rd Great-Grandfather had wanderlust, packing up the family and moving on after a while for different opportunities. While my family arrived in Ohio via horse-drawn wagon, they left for Kansas, by way of Missouri, by the railroad in 1884, arriving just in time for the last of the great cattle drives. The cattle drives came to an end, along with the mechanical era, when the coal-powered railroad spread out across the lands providing everyone access to fossil-fuel driven transportation.

3

Human History is Energy History

> *"Humanity was in an abysmal state before widespread use of fossil fuels and the products manufactured from petroleum derivatives; and nuclear-generated electricity when self-government, free speech, and necessities we take for granted did not exist."[81]*
>
> Stein & Royal, **JUST GREEN ELECTRICITY**

If there is one thing that seemingly everyone agrees upon, the earth is falling apart, and we must all do something to fix it. When asked, "All things considered, do you think the world is getting better or worse?" only 6% of Americans and 3-4% of Europeans answered that it was getting better. In a 2017 survey across many countries, only 35% said the future would be better than the present. Overall, the people taking the survey had the facts wrong 80% of the time on global poverty, 61% on child mortality rates, and 90% of the time on the world's current state.[82] However, it is rare in human history that the future has not been better. Humanity has experienced many changes to how life is lived, changes to the basics of food, water, warmth, and how society interacted. Are there lessons to be learned from those transitions?

The **Green Solution** represents one implementation of the evolving plan in which Nations call for taking a new step forward into the reality of how man will use energy to his benefit. While transitioning to new energy realms is not new in history, this one will be the fastest ever attempted. It is the first time that it is an intentional act of man and not having naturally evolved due to changing environments or discoveries.

How and what energy man uses must be viewed as phases that last for a block of time in which a manner of behavior will dominate society. Generally, these phases are defined by two basic human behaviors: transportation and heating/cooking methods. In both areas, man has evolved from manual to mechanized over time. One must be realistic and realize that energy use will always vary across a society based on the wealthiest and most impoverished populations. The rich can live in one phase while the poor live in another. Further, it is clear that nations worldwide sit at different places on the energy use spectrum, with some parts of the third world far behind the developed nations.

The history of primary energy consumption in the United States has transitioned from wood to coal to fossil fuels. This pattern is not unlike that experienced in the rest of the world, as one nation's technical achievements have always moved to impact other nations' developments which continue to build on the ideas.

As Ronald Bailey has noted, "With regard to most of human history, there has been precious little in the way of development. The vast majority of people lived and died in humanity's natural state of disease-ridden poverty and pervasive ignorance."[83] Human history without a clean, reliable, and safe energy source for man's basic needs has been chiefly a fairly hopeless existence. For most of history, unless you were a soldier, explorer, or on religious pilgrimages, it was rare to leave the district in which you were born. Spending an entire lifetime within a 25- or 50-mile radius was more typical than not. With low life expectancies, limited travel, and limited food sources, life in the past was pleasant only for the wealthy and the elite.

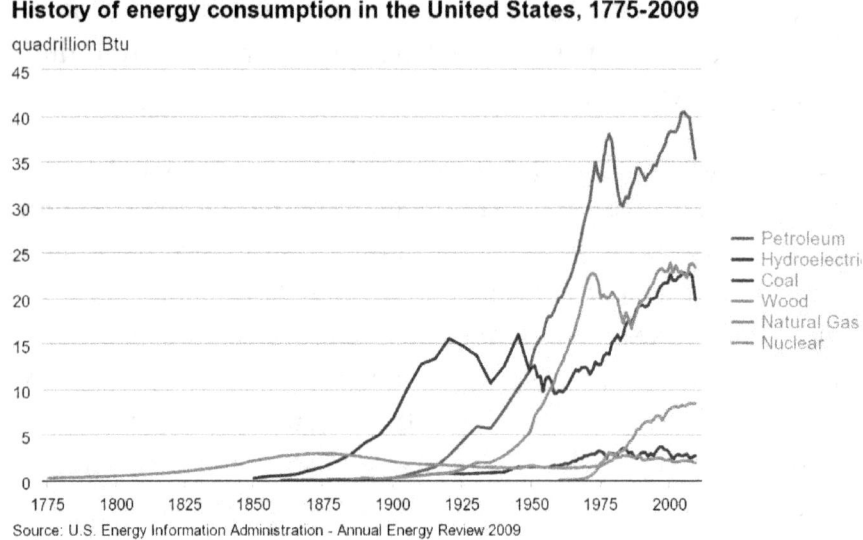

Figure 12 History of Energy Consumption in the U.S., 1775-2009

Cesare Marchetti, an Italian nuclear physicist, created models in the 1970s to explain when and how the world would transfer from one source of power to another. He called these points "energy transitions." He

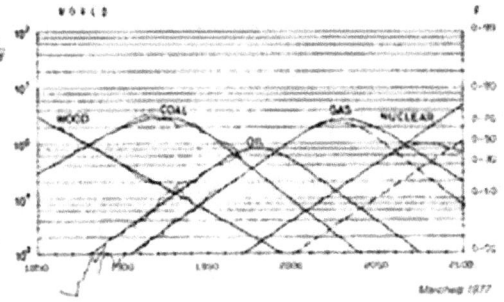

Figure 13 Energy Transitions

revealed how man has historically worked himself from sources of lesser power density towards higher power density sources through his studies. As the technology develops for each higher density source, the amount of energy available per cost drops substantially and requires less effort on the part of men.

As the developing countries undergo advancements in their energy transitions, different parts of the economy may be experiencing different phases at any particular point in time, especially between the urban centers and the agricultural communities. As French historian Fernand Braudel noted, "In no society have all regions and all parts of the population developed equally."

Energy Transition #1: Of Man Alone

The earliest generations of man were the hunter-gatherers whose primary energy source was their own muscular strength. He had fire for heating and cooking, but it was the strength of man that chopped and collected the wood and tended the fire. Progress was limited by the extent of that which a man could physically do. Access to tools and shelter was possible in whatever he could handcraft from the natural materials around him. With such limited resources available, he could do little more than hunt and gather. Hoping that he could collect enough to feed his family on any given day and provide enough to restore the strength that he would need the next day. There was little a man could do to multiply his effort on his own. So by assembling in communities or even capturing slaves, the margins to survival could improve for all.

Survival proved too difficult, and, in general, humans moved beyond this manner of life long ago. However, there are still communities on the planet today that reside in this mode. Roughly 100 uncontacted communities[84] represent something less than a quarter of a million in population, mainly in the Amazon, African forests, and a few remote islands. These uncontacted people frequently have limited availability of modern tools and weapons unless they are found, stolen, or traded.

Is progress required? Organizations like Survival International (survivalinternational.org) argue that we should recognize their right to self-determination and view them as "our contemporaries and a vitally important part of humankind's diversity." Their proposed plans for these people include "protect their lands and ensure their right to remain uncontacted is respected." A recent United Nations report calls for states to "redouble efforts to protect the territories" of the tribes. This tribal lifestyle should not be viewed as a thousand-year-old Stone Age lifestyle but one that has grown and matured in a different direction.

Like that of people everywhere, these tribal groups have been continually evolving and adapting to the environment in which they live, an environment that is abundant in resources, food, and game. This lifestyle could not be supported in other settings, so new living methods evolved, which moved humanity towards another energy transition.

Energy Transition #2: Man Gains a Companion

Soon after taking up agriculture somewhere around 5000BC, humanity began to domesticate animals both for food and to use as a work companion. A man's daily accomplishments increased when the domesticated animal came alongside, as the workload could be what both the man and beast could achieve. Heavier loads could be carried across long distances with this new arrangement, loads which would have taken several trips by a man to accomplish. Man, slave, and beast could tend more farmland, build more buildings and herd more animals than the man could do alone.

In this regime, cooking and heating were still based on the ability to burn a substance. In many places, this was primarily wood products but also included other plant material and animal dung. In chapter 19 of her 1940 book, **THE LONG WINTER**, Laura Ingalls Wilder describes the process of twisting hay into the hay sticks that her family burned to survive the severe winter of 1880-1881 while they lived in the Dakota Territory.[85] They were reduced to hay burning as a fuel source due to the area's lack of trees. A technique still used in many parts of the world today.

While the uncontacted tribes are small and isolated, half of humanity lives in the fifty developing countries that still rely heavily on draft animals. According to the UN Food and Agriculture Organization (FAO), the developing countries use some 400 million draft animals, including 139 million water buffaloes and 20 million camels,[86] to work on 52% of the farmland. There are an estimated 25 million animal-drawn carts,[87] of which some are of a traditional design of wooden wheels and frames while others use scavenged steel frames and truck tires.

In the developed world, when the demand for manual labor increases, the need is met by those from the lesser developed nations. However, where there is no one lower on the economic ladder, the ugly nature of slavery tends to arise. Today, the estimated number of slaves being held ranges from 38 to 46 million[88], with some 25 million of those still being held for forced labor in domestic work, construction, and agriculture. Incidents of forced labor are most predominant in those areas where the power of man and beast are still the dominant energy sources.

Current societies existing in this energy phase have the lowest life expectancies (around 51 years), and the rates of child mortality and poverty are the highest. In 2000, The World Health Organization (WHO) reported that "Approximately half the world's population and up to 90% of rural households in developing countries still rely on unprocessed biomass fuels in the form of wood, dung and crop residues. These are typically burnt indoors in open fires or poorly functioning stoves. As a result, there are high levels of air pollution, to which women, especially those responsible for cooking, and their young children, are most heavily exposed."[89] By 2020, the number using such biomass for cooking was still at 2.5 billion.

While this energy phase reflects some portions of the least developed nations today, other parts, mainly urban areas, of those societies as well as the developed nations have undergone further energy transitions.

Energy Transition #3: The Time of Mechanics

It was the development of mechanics that represented the first step into the industrial revolution. Without the technology provided by the revolution in mechanics and mechanisms, the industrial revolution could never have succeeded as it did. The interim time of mechanics brings fundamental changes to humanity, an end to the slave trade, and fuel introduction.

This was a time represented by a collection of means by which work could be improved without using fuels to accomplish the tasks. Using wind and water to power mills and factories such that multiplied the work that the urban residents could achieve. While in the fields, productivity gains brought by all the mechanisms reduced the demand for physical labor, which ultimately eliminated the need for a slave workforce. The transition begins with the 300 labor hours to create those 100 bushels of wheat but will end with requiring only 75 hours to complete the same tasks.

It was mechanics and the introduction of fuels that brought an end to the economic institution of American slavery. However, it would still require a civil war to bring an end to its social institution.

Energy Transition #4: Man uses Primary Fuels

Primary Fuel: Fuels that are found in nature and can be extracted, captured, cleaned, or graded without any sort of energy conversion or transformation process.[90]

Primary fuels are the power of plants, whether recently grown plants as in wood and hay or plants long since grown and buried in the form of coal. These fuels are those that can be collected and used directly from nature, whether in the sense of cutting down a tree or digging underground for coal.

Historically, using coal collected from the earth's surface dates back a millennium in China and was heavily used in the Roman empire. Still, coal as a fuel was pretty much lost and did not resurface until the 13th century. Throughout the Renaissance, the mining industry began the process of retrieving coal from underground.

Throughout the 18th century, the steam engine was developed and underwent many improvements, rapidly proceeding from a clever idea to a handy gadget. The steam engine's uniqueness was the ability to take fuel (wood and later coal) and convert them to functional mechanical workhorses. The steam engine's initial application was to pump water out of mines, an activity previously done by men at the pump handles or animals walking in circles to operate the pumps. Through the engine's use, the wood or coal energy could free men from the tedious tasks of manning the pumps. Without operating pumps, the lower levels of mines, especially in England and Wales, would flood, endangering the men below and hampering production. Thus the pumps could never stop running.

With the steam engine's advancement, the engine could become powerful enough to move cars along a set of tracks, leading to railroad development. Up until the 1870s, American locomotives ran primarily with wood for their boilers. The energy obtained from a load of wood was limited, thus restricting the distance that trains could run before needing to collect another load. The absence of trees across the Midwest meant that the next load of wood could not be sourced locally and had to be shipped from the east. The development of a transcontinental railroad would have been difficult based purely on wood as fuel. The switch to the higher energy density fuel of coal allowed the locomotives to run farther and the stocks of fuel across the Midwest to be often sourced locally and maintained more efficiently.

As with most technology, the advancements in the steam engine made them smaller, cheaper, and easier to maintain. Now, rather than being restricted by needing access to waterpower, textile and other factories could open anywhere, using coal to power those drive belts. The unfortunate downside was that without the need for physical strength, child labor tended to increase in these facilities.

The Mayflower took 66 days under sail to reach America averaging just 2 miles per hour, basically a leisurely walking pace. The 102 passengers spent the days in the extreme discomfort of being cold, underfed, and often trapped below. It was a miserable passage. With the advent of primary fuels, the SS Great Western provided its 128 passengers with 1st class cabins in the 15-day journey of the first steamship crossing in 1838. The transatlantic record for a primary fueled ship is held by the SS United States and its 1,928 passengers and 900 crew of 3 days, 10 hours, and 40 minutes in 1952, an average speed of 39 miles per hour.

With primary fuels, people and goods could be moved rapidly and at great distances. However, the skies of the industrial centers soon darkened as the use of coal for running factories, powering trains, and heating homes grew. The pollution grew with the introduction of each new engine, blanketing cities in both Europe and the New World. The choking smog would continue until an energy source could be developed where a central powerplant could provide clean energy for all.

Energy Transition #5: Man Creates Secondary Fuels

Secondary Fuel: Not found as a natural resource, secondary fuels are derived from primary fuel or fuels through chemical or physical processes.[91]

Not to disrespect the effort to cut down trees or dig underground to collect coal, but secondary fuels are those created when you take a primary fuel source and process it to make a better, handier, safer fuel with higher energy density.

Secondary fuels principally are represented by the derivatives of crops(ethanol), wood (Charcoal), Coal and Natural Gas (Electricity), Petroleum (Gasoline, Kerosene, Propane, Diesel), and Uranium ore (Nuclear fuel rods). Each fuel derivative functions as being an improvement over the primary fuel from which it came. After making an exception for the nuclear fuels with their own unique considerations, each of the fuels also becomes much more consumer-friendly and usable in smaller quantities.

The personalization of transportation becomes possible within the realm of secondary fuels. Steam engines generated helpful work from the primary fuels of wood and coal; however, these engines were large, complex, and tricky to manage and therefore much more suited for locomotives, ships, and other large means of transport where specialist engineers could watch over the engines. While the Stanley Steamer and other steam engine cars were developed, the practicality of the internal combustion engine burning the secondary fuels made it possible for everyone to have personal transportation devices.

Table 2 Automotive Registrations by Fuel Type

Fuel Type	United States	Europe
Gasoline	78%	55%
Diesel	14%	38%
Electric	8%	5%
LNG	<1%	1.5%

With internal combustion engines developed for automobiles came an engine with enough power for its weight to enable flight. With flight, the first transatlantic crossing lowered the time required to 33½ hours. Then, the Concorde, powered with petroleum-derived jet fuel, the transatlantic record was reduced to just 2 hours and 53 minutes—a time just 0.2% of the 66-day journey for the Mayflower passengers.

> *"Electricity is a modern necessity of life"* – President Franklin Roosevelt, at the Rural Electrification Administration celebration, 1938.

Electricity was slow in development in that while the idea and power of electricity was understood in the 1750s, it takes nearly 150 years for the concept to become a reality. A source of energy that could be created on-demand, delivered on-demand, and used in a productive manner by everyone. From the first electrical grid that Thomas Edison installed in 1882, the availability of electricity spread worldwide, reaching 87% of worldwide households by 2016. Beginning with lighting, electricity found uses in household appliances, a market which has grown to a $133 billion business. It has been shown that the product which provides the most freedom for women is the washing machine. Without washing machines and depending on the availability of soaps, a family's laundry would require an enormous amount of physical labor and could take over 15 hours to complete. When society has developed to the point where families have washing machines, then women tend to be freer. They work outside the home at three times the rate, the girls are 5 times more likely to receive a full education, birth rates decline by half, and women receive equal rights. To have washing machines in the home, society needs two things: enough wealth for the average family to afford a machine and stable and reliable electricity to give the confidence to use the machine.

The era of nuclear power comes into its own after the second world war. Uranium is considered a secondary fuel as the uranium ore coming from the ground must be processed to create concentrated fuel rods. Depending on the reactor type, this uranium may need to be enriched with a higher percentage of U235 isotopes than naturally occur or even conversion to plutonium-239. From powering cities to driving naval vessels, the nuclear reactor has generally been safe and pollution-free during its 70-year reign. While many people have been hurt from uncontrolled radiation incidents, of the three major nuclear reactor accidents, no one died from Three Mile Island or Fukushima, and 2 facility workers and 28 firefighters perished at Chernobyl. While tragic, this represents the lowest fatality rate of any energy source.

While coal flooded the skies with dark soot, secondary fuels reduced the pollution rate, and the cities slowly cleared. Yet, carbon emissions are still very much a part of burning secondary fuels. Only by going beyond secondary fuels can atmospheric carbon be reduced.

The Final Energy Transition: Carbon Emission Free Life

The final energy transition to a life without carbon sources would be the first to be preplanned to happen. As Fischer-Kowalski and Harberl commented, "this transition some time ahead is inevitable, due to the exhaustibility of fossil fuels. How far ahead and whether the transition happens inadvertently or by deliberate planning and intervention is an open question."[92] The accomplishment of becoming truly carbon-free requires, "halving of the metabolic rates on the part of the industrial nations... metabolic rates not much higher than at the beginning of the twentieth century." The effort will be extensive as the "transition to a more sustainable state implies a major transformation, comparable in scale to the great transformations in history such as the Neolithic or the Industrial Revolution."[93]

A decarbonized society can take two approaches: generate no carbon emissions or capture those emissions which it does generate. Moving all power sources to the electrical grid, so-called "electrify everything," moves the source of emissions to major electrical power stations, where the carbon could be captured and sequestered. However, merely capturing the emitted carbon fails against the standard of sustainable energy, considering that fossil fuels must at some point run out. However, it would represent a first step towards the ultimate goal.

The environmental news company, EcoWatch, ran a story on their website in September of 2015 toting the 96 cities that were "quitting fossil fuels and moving toward 100% renewable energy."[94] As of the article's date, the cities of Aspen, Santa Monica, San Francisco, and Stockholm had reached that coveted status of being completely fossil-fuel free. Later in the article, they admit that they "decarbonize[d] their electricity supply" and did not eliminate all fossil fuels in the city.

A decarbonized energy network would necessarily consist of wind, solar, hydroelectric, and geothermal renewable energy sources and likely to include nuclear power as well. Renewables like wind and solar have been used for centuries in the form of windmills driving pumps and providing mechanical drive, while solar arrangements have provided heating, especially as a source of hot water.

The carbon-free transition presents three primary obstacles which the engineers will need to overcome to make the change possible, demand reduction, overcoming intermittency, and eliminate regional deficiencies.

A nation's metabolic rate is a measure of the energy used per unit of gross domestic product (GDP) generated by the nation. Since the carbon-free sources tend to be of lower energy density than secondary fuels, it will necessitate keeping the amount of required energy under control through increased efficiency of the energy used and a reduction in society's power demand.

The second obstacle is to deal with the fact that solar and wind only work when the sun shines, and the wind blows. This issue can be addressed either by demand reduction or storage. If attacking the problem from a storage consideration, enough capacity would need to be built such that power demand can be met while still having enough left over to charge various storage methods, whether physical, mechanical, or chemical. Otherwise, the reduction methods require that when the sun doesn't shine, or the wind doesn't blow, there is no demand for electricity.

Finally, an energy plan that depends on natural resources like sun, wind, and flowing rivers must adapt to regions where they are less available. Solar is effective in the south, particularly in the summer, but is considerably less effective in the northern winters. Likewise, wind power varies based on geography, with the coastal areas having more consistent winds. Any energy plan must have a plan to provide power for those regions without rivers, are too far north for solar, and have limited wind power available. The Solutions Project presents such a plan having specified the wind, solar, and hydropower generated by each region. This analysis shows that Texas generates the most wind power, the southwest generates the solar energy, and the west coast provides the hydropower, while high voltage lines must cross the nation to get the power to the east coast where demand is higher than the ability to generate it.

Across the globe, communities exist in each of the six energy transitions, each tuned to the local needs, demands, and their technological achievements. It is the push of the **Green Solution** to move everyone on the planet to the final transition, regardless of where they currently operate. There is, however, an open question of how to address those communities that operate in phases other than those of the highest technology enjoyed by the developed nations. While the Haves live in one energy transition, the Have-Nots live in a number of different circumstances. Can these less developed communities leapfrog certain phases and move directly to the final transition, and what would be the consequences of such an attempt?

A Glimpse into Life in the Industrial Age

At the time of the American Revolution, 97% of the population labored on farms. When the industrial age took hold of mainstream America, it significantly impacted how their lives were lived. By 1900, the United States population that lived on farms had dropped to 38%, but that number was significant enough to include all eight sets of my great-great-grandparents. In those years, my ancestors lived on farms in Kansas, Missouri, South Dakota, and Indiana. The Department of Agriculture estimates that the total farm population peaked in 1916, and since then, the farm population has fallen to about 2.6 million people or 1.3% of the population. Those 100 bushels of wheat, which started off requiring 300 hours of hard labor, now, with the aid of modern farm machinery, need just 10 hours, most of which is spent in the cab of an often air-conditioned tractor.

Falling in line with the trend, my own family moved off the farm in the very next generation. Of my four great-grandfathers, three made livings related to the fossil fuel industry, as a gas station owner, railroad engineer, and coal miner. Even though one of my great-grandfathers continued to work on the farm, none of his children followed suit, becoming involved with education and sales. Like so many families of their era, it took just two generations for my ancestors to move away from agricultural work in favor of life in the city and work in the newly industrialized economy.

As a result of moving off the farm, my family became wealthier and healthier. Like a majority of humanity throughout history, they lived on and farmed land owned by others. But after a few short years, they owned their own homes that were warmed by natural gas rather than wood. They read by electric lamps rather than candles and drove their own gasoline-powered automobiles rather than horse-drawn carts.

Throughout human history, individual ownership of land and property was rare and typically reserved for those born into the privileged class. With the industrial revolution, this all changed, with those who left the farms and moved into the urban/suburban communities gaining wealth never before possible.

By the 1950s, with a third of Americans living in suburbia, 60% owned homes and 75% automobiles. In 1957, out of all the homes wired for electricity throughout the country, 96% had a refrigerator, 87% an electric clothes washer, 81% a television, 67% a vacuum cleaner, 18% a freezer, 12% an electric or gas clothes dryer, and 8% air conditioning. All of which was unheard of just one century earlier. Even the White House did not get air conditioning until 1933.

The city dwellers were rapidly becoming healthier and wealthier. Their life expectancies, especially for males, improved faster than their rural- and farm-dwelling counterparts. In every measurable aspect, life had become better than at any time in recorded history.

Will the advancements of this age be the last of such advancements? That will depend on the decisions of the future.

4

The Great Divide

> *"Tackling climate change is a complicated undertaking, to say the least. But here's a good rule of thumb for how to get started:*
>
> *Electrify everything."*[95]
>
> Dave Roberts, VOX

One of the significant faults in the entire Climate Change / Fossil Fuel discussion is the problem's oversimplification. Lumping all fossil fuels together without considering their actual use leads to discussions that can at times be completely non-sensical. Making the generalized statement of "Ban all Fossil Fuels" fails to reflect the reality of how those fuels are used.

Not all fossil fuels are created equal nor have equal applications. Any trip to the gas pumps these days demonstrates this nicely in a microcosm. The typical gas pump has two to three nozzles from which to purchase a product. These nozzles allow for the delivery of gasoline, diesel, and possibly an ethanol-based fuel, like E85. Each fuel uniquely fulfills a limited number of applications. Often vehicle manufacturers will use stickers that will

Figure 14 Multi-output Gas Pump

indicate exactly which nozzle to use. The government regulators have looked after us, putting different size nozzles and different colors on the handles, making it difficult to put the wrong fuel in your car. Ignore their instructions, and disaster is usually on the horizon when the wrong fuel is used in an engine that is not designed for it. Put diesel into your gasoline engine, and you won't get very far down the road before you are calling for help from the shoulder.

Likewise, fossil fuel raw materials, namely Coal, Natural Gas, and Petroleum, have different destinations as to their use. One would never consider putting coal into the tank of an automobile and expect it to work. Yet, some plans and their accompanying comments would lead one to believe that it might actually work. To be clear, coal-derived gas and oil products do exist in small amounts, but they do not play a large part in the overall discussion of fossil fuel use and distribution.

It is helpful to review how energy is used in our current environment. Generally, energy is used in two ways, either as electricity from the electrical grid or directly from petroleum derivatives, such as gasoline, propane,, diesel, aviation fuel, or natural gas. The current distribution is that 18% of the world's energy is in the form of electricity.[96] Within the United States, electricity represents about 17% of all energy consumed.[97]

By end-use, electrical energy usage was such that residential consumption is 38%, commercial 36%, industrial 25%, and transportation at 0.2%.[98] In the table below, the total consumption of energy (in Quadrillion BTUs) is listed for the different areas of end-use, which are residential (homes and apartments), commercial (stores and offices), industrial (manufacturing), and transportation (planes, trains, and automobiles). The final entry is for "system losses," which helps balance the account for all the energy used. When converting any fuel, whether natural gas or coal, to electricity, the input fuel energy will always exceed the outgoing electricity's energy. This loss is a property of thermodynamics and while it can be minimized, it cannot be eliminated. When combined with the energy losses due to the long transmission lines, experienced as an audible hum from high voltage lines or static on your car radio, and the power required to run the systems that generate the power, the total losses are about 65% of the input energy. The end result is that 26% of all energy in the United States is used to generate electricity.

Table 3 Total Energy Consumption by Area of Use in Quad-BTUs

Area of End Use	Electricity (from Coal, Natural Gas, Nuclear, Renewables)	Non-Electrical (Petroleum Derivatives + Natural Gas)
Residential	4.9	7.0
Industrial	3.2	23.1
Transportation	0	28.2
Commercial	4.6	4.8
System Losses	24.3[3]	

When planning how to move away from fossil fuels, it is not only essential to understand that energy is divided between electrical and non-electrical uses but also what applications for which the energy is used. Depending on the application, there may be better energy sources to use. For example, natural gas can indeed be used for lighting homes, but electricity might be a better choice. On the other hand, using oil for lubrication is better than trying to use electricity. Table 4 shows some examples of how the different end-use areas use the various forms of energy.

---

[3] Defined as transmission and distribution losses (7%), In plant use (5%) and thermodynamic loss as part of input to output conversion. See https://www.eia.gov/totalenergy/data/monthly/ Section 2 Note 1. Electrical System Energy Losses.

Table 4 Current End-Use Distribution

Area of End Use	Electricity (from Coal, Natural Gas, Nuclear, Renewables)	Non-Electrical (Petroleum Derivatives + Natural Gas)
Residential	Lighting Heating / Cooling Cooking Appliances (refrigeration, laundry) Entertainment (TV, Internet, Phone) Technology Security	Heat Cooking Hot Water
Commercial	Lighting Technology Point of Sale Services Cooling Refrigeration	Heating Cooking Hot Water
Industrial	Lighting Technology Cooling	Heating / Cooling Process Heating Chemical Feedstocks Petrochemical
Transportation	Electric Automobiles (1%) Subway / Light Rails	Automotive Trucks and Transport Trains Aircraft Lubrication
Support of Energy Generation	Lighting Technology Pumps	Lubrication

A couple of the more common items on the list were the energy usage for heating our environments and cooking our food. Since they are listed under both energy categories, it would be good to understand how the uses are broken down. In America, natural gas dominates how we heat our homes, with electricity, namely heat pumps, being second. While we have more electric cooktops than we have natural gas ones.

Home Heating[99]:

Natural Gas: 48%
Electricity: 37%
Fuel Oil / Kerosene/ LPG: 12%
Other (wood): 3%

Cooking[100]:

Electricity: 63%
Natural Gas: 35%
Other (wood): 2%

Note: LPG: Liquid Petroleum Gas

In the second Presidential debate, President Biden made that troubling statement, "I would transition away from the oil industry …It has to be replaced by renewable energy over time." Whenever discussing renewable energy, we must be careful to understand what is included in that category. Are we talking wind and solar? Maybe also hydroelectric, or perhaps also crop-based ethanol? As we learned from the tables above, the "oil industry" and "renewable energy" (except for ethanol-based fuels) are primarily in different columns. What the President meant by the comment, and what most environmentalists mean by their proposals, is that they intend to move items from one column to the other. The implication of such a statement is that such a move requires the development and installation of new equipment and sometimes new technologies. While it is true that an apple can replace an orange as a fruit, it does not mean that it has the same function and straight-up replaceable, especially if you ask the apple growers of Washington or the orange growers of Florida.

The **Green Solution's** final result should transform the end-use chart by moving many elements from the secondary fuels column to the electricity column. That, of course, transfers the demand for the energy from petroleum products to whatever source is used to generate the electricity. With this proposal, the current usage of 17% of our energy in the form of electricity would necessarily rise to over 70%. If the table is modified considering the plan to "electrify everything," especially the household and transportation sectors, it would be close to the table below.

Table 5 Post Green Solution End-Use Distribution

Area of End Use	Electricity (from Coal, Natural Gas, Nuclear, Renewables)	Non-Electrical (Petroleum Derivatives + Natural Gas)
Residential	Lighting Heating / Cooling Hot Water Cooking Appliances (refrigeration, laundry) Entertainment (TV, Internet, Phone) Technology Security	
Commercial	Lighting Heating/Cooling Cooking Hot Water Technology Point of Sale Services Refrigeration	
Industrial	Lighting Technology Heating / Cooling Process Heating	Chemical Feedstocks Petrochemical
Transportation	Automotive Trucks and Transport Trains Subway / Light Rails	Aircraft Lubrication
Support of Energy Generation	Lighting Technology Pumps	Lubrication

A great chasm separates those parts of our economy which are powered by electricity and those elements which are powered by the primary and secondary fossil fuels. There are definitely applications that would be best in the electrical column and those best in the non-electrical one. Innovators have used energy from both columns to attempt to solve many problems. The initial development of the automobile included vehicles powered by gasoline, diesel, batteries, and steam, and the market determined that gasoline provided the best solution. Likewise, it was once possible to purchase coffee pots and toasters that ran off the natural gas lighting, but they lost out to the electrical solutions. Currently, there are clothing irons on the market which are heated by electricity, gas, or coal, but the market has determined that electricity is the best solution, except where electricity is not reliable. In those places, the coal and gas heated versions still sell. Repeatedly, the market has focused applications to the best energy solutions.

At first glance, it would appear that some products like the Hybrid Vehicle would provide a bridging technology between the two columns, but not so much. A hybrid uses electronic technology to enhance the fuel efficiency of the gasoline-powered vehicle. Consider this, without the battery, the car is a less efficient version of the same vehicle. In contrast, without the engine, the car is a severely compromised version with significantly less range and functionality. Although some hybrid vehicles can now be recharged by a plug-in, a majority can only be recharged when operated with the gasoline engine.

As we move forward to evaluate the benefits and risks of fossil fuels and their replacements, keep in mind the lesson of the great divide. Those who understand the issue, understand the divide, and those who lack understanding, act as if there is no divide about which to worry. All of which can lead to policies with severe negative consequences.

5

Lessons in a Barrel

"Use Old Dinosaurs, not New Trees."

Jon Huntsman, Sr. CEO of Huntsman Chemicals

He was big, he was green, and he covered the roadway landscapes of my childhood. He was the brontosaurus of Sinclair Oil Corporation. His name was Dino. Sometimes on the lawn in front of the station, and sometimes up above the gas pumps, he varied in size and location, but he was always there. He was there on the advertising signs, and he existed as tiny plastic Dinos and large blow-up inflatable ones. Even today, a tour of most antique / resale shops can usually dig up a Dino or two. Across middle America, he still makes appearances alongside the roadway. While in New York, he annually makes appearances as a balloon during the Macy's Thanksgiving day parade.

Figure 15 Sinclair Logo

It began when Sinclair sponsored the dinosaur exhibit at the 1933 Chicago World's Fair, in which Dino helped "promote lubricants refined from crude oil believed to have formed when dinosaurs roamed the earth."[101] Dino is classic advertising genius, one that we can never forget and recognize at a glance, even getting cameo appearances in the Toy Story franchise of movies.

The link between the brontosaurus and oil became so iconic that when in 1991, the ABC network brought to prime time the show Dinosaurs!, a show about a family living some 60 million years ago, he worked for the oil company WeSaySo. And his surname was, you guessed it, Sinclair.

The message was clear, oil came from dinosaurs, and after all, they were called fossil fuels. And what we know of dinosaurs are from the fossils. The moniker, penned in 1759 by a German chemist named Caspar Neumann[102], was meant to indicate that these fuels were buried and found in the earth. All this business about dinosaurs is misleading. The material that decomposed and underwent the chemical changes that have created oil was primarily plants, phytoplankton, and zooplankton. These plants utilized CO_2, water, and solar energy to grow, building the complex carbon molecules they left behind.

Today, the ubiquitous barrel of oil has become the driving force of everything in our world. It is the basis of all movement, comfort, and finance. Ask the average person on the street, and they will equate oil with transportation, but it is far more than that. Although roughly half of the products associated with a barrel of oil are used for the aviation, diesel, and gasoline fuels driving our transportation needs, the remainder is used for base chemicals used to manufacture the petroleum derivatives to support our lifestyles and desire for comfort.[103]

To intelligently discuss the literally thousands of products that are derived from that barrel of petroleum, it is necessary to fully understand how the oil in the ground becomes those wonder products. What starts life as a dense black liquid buried deep underground eventually comes into our lives as medicine, fabric, soaps, plastics, electronics, and structural materials.

> *"Buy soon! Before this wonderful product is depleted from Nature's Laboratory."*
>
> *An 1855 advertisement for Kier's Rock Oil*

The term "dinosaur" dates to 1841, when Sir Richard Owen first coined the term during a talk entitled "Report on British Fossil Reptiles." The term is not as old as many would think and aligns closely with the history of petroleum. As the new terminologies of dinosaurs and petroleum arrived together, so, in the public's mind, they have been forever linked.

Even as Sir Owen talked about dinosaurs in Britain, Samuel Kier was in Pittsburgh marketing his "Kier's Petroleum" as a liniment for burns and "cure for all that ails you." "Get it quickly," they advised as the amount of this petroleum would surely run out.[104] Selling in an 8 oz. jar at the cost of 50 cents, the black product represented the waste from when his salt wells became fouled with the stuff. Kier operated a salt business where water was injected into underground salt veins and pumped back out to retrieve the dissolved salt. Unfortunate for him but fortunately for us, he found that the fouling product was burning in the streams into which it had been dumped. The first oil well had not yet been drilled, but the products' marketers were already threatening the petroleum era's end. It seems that some things never change.

While still claiming it was running short, Kier kept collecting more and more of the oil. More than he could sell for medicinal purposes, so tinkering began with distilling to create "carbon oil" for other purposes. The hunt for the magical liquid was on.

Colonel Edwin Drake, who wasn't really a colonel at all, found backing and drilled the first production oil well in 1859 in Pennsylvania, a region that would dominate the petroleum industry until the Spindletop well was drilled in Texas in 1901.

What is in a barrel?

In order to fully understand the implications of the fossil fuel industry, it is worth taking some time to understand what exactly is in a barrel of oil. Crude oil is a soup of molecular compounds of varying carbon counts. Therefore, differing weights lead to the compounds having different attributes and uses. It is the refining process that separates these compounds into their many output streams.

If one were to go to their local wholesale club store and purchase a box of individualized bags of potato chips as are used in lunches, the package would likely contain an assortment of flavors, maybe Sea Salt, Green Onion, and Barbecue.

Figure 16 Popchip Selection

What if you were not a fan of Green Onion chips? Could you get a box without those? Well, not without opening and sorting them out, which would be frowned upon by most stores. No, one must purchase the entire box of 30 bags, and every package gets the same variety, 12 bags of Sea Salt, 12 bags of barbecue, and 6 bags of Green Onion. Even if you are a real fan of Green Onion, you are stuck with precisely six bags per box, no more and no less.

Likewise, a standard barrel of oil is a measure of 42 gallons of pre-refined crude. After the refinement processing, the resulting material would be upwards of 45 gallons of final products. Although the exact numbers vary depending on the oil source (not all oil is created equal), a barrel provides various compounds in a standard ratio with a slight variance. The output from the refinery of a standard barrel of oil makes up about 47% Gasoline, 33% other fuels (diesel, aviation, kerosene), 18% various petrochemicals, some natural gas, and some Asphalt[105]. The product types extracted are determined by the compounds' chemical nature within the original oil soup. It is not possible to change the ratio or nature of the compounds unless changing the source of the crude. What you have is what you get. Converting heavy asphalt material to be lighter jet fuel would require a lot more processing than just the refining process.

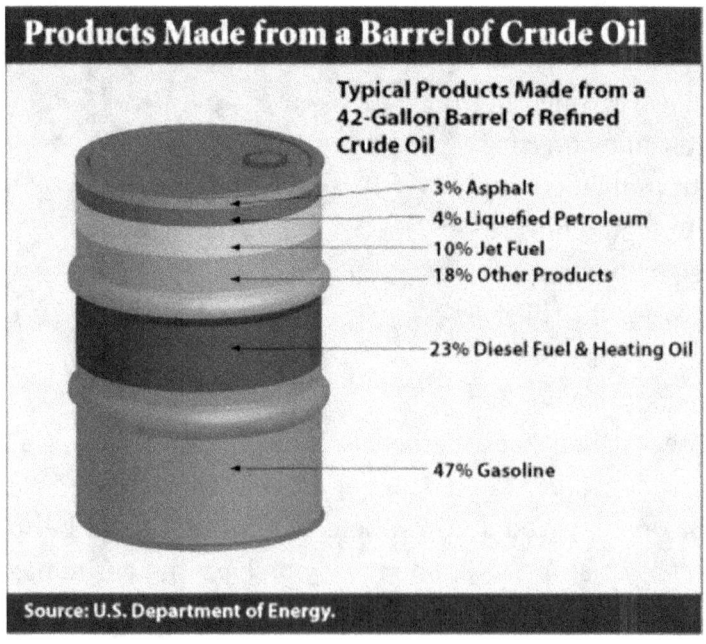

Figure 17 Products made from a Barrel of Crude Oil

Selecting crude from different sources can be much like choosing different brands of potato chip boxes. While the product mix can change and even new products can be available, even with the new brand, Cheddar and Cheetos are fixed in ratio with plain, green onion, and barbecue.

Barrels of Oil and boxes of potato chip bags have very much in common; either you get everything in a fixed ratio, or you get nothing at all. It is a bundle deal that you cannot break without taking any unwanted product and recycling, giving it away, or sending it into the waste stream.

Figure 18 Baked Mix Chips

Starting from the early days of discovery, a refining process has been developed over time to extract more and more differentiated products from the crude oil compound soup. There are organic (containing carbon) and inorganic (carbon-free) compounds within the soup, including several thousand different hydrocarbon compounds.[106] Sulfur, Nitrogen, Oxygen, and Hydrogen exist in mass quantities, along with lesser amounts of various metals. The refining process sorts the compounds based on the number of carbon atoms in the molecule. As the diagram that follows shows, at the top of the stack is the light gases with 1 to 4 carbon atoms, while at the bottom of the stack are heavy tars and asphalts with 70 or more carbon atoms. As these products vary by chemical compound, it would take a chemical process to combine or break apart the molecules to create a different product from the original one, all of which takes energy and other raw materials. Generally, compounds with less than 5 carbon atoms exist as gases, while counts 5 to 17 are liquids, and greater than 17 carbon atoms come out as wax-like solids.[107]

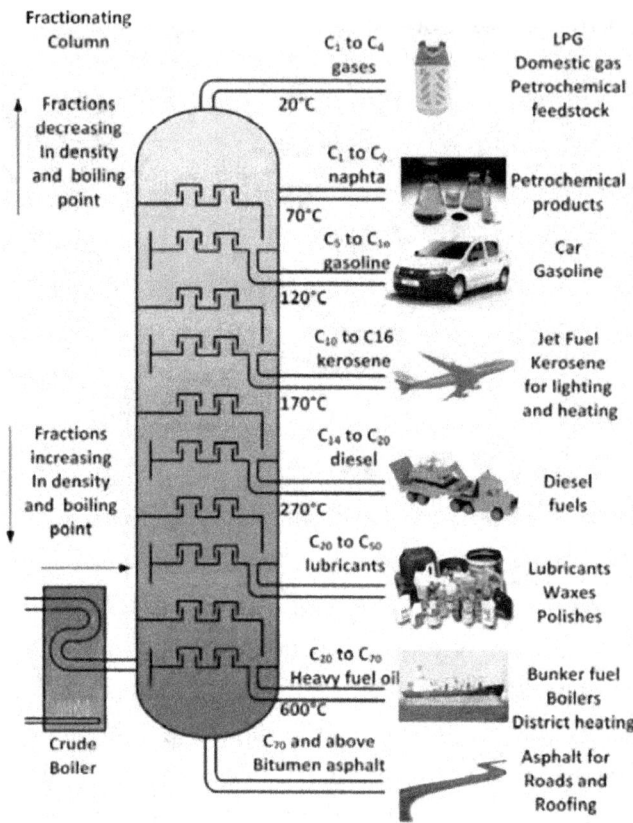

Figure 19 Refining Outputs

Knowing what is in a barrel of oil is critically important to know so that we do not fall victim to the over-simplification of the process. Each of the categories of output from the refinery process develops its own industries.

The industries that have arisen from the availability of fossil fuels can be thought of as technology suites. The definition of a suite is that it is a collection of things to be used together. A technology suite is the collection of technologies and industries that makes it all possible. Considering that oil provides a certain amount of gasoline per barrel, this has enabled the creation of a technology suite to provide for the use of that gasoline. In this case, the suite consists of the obvious like the drillers, refineries, delivery, and sales mechanisms for the gas and the slightly less obvious in the vehicle design and manufacture, sales, and delivery. Additionally, it includes entities and employees extending from car service centers, parts stores, accessory manufacturers, auto loan specialists, auto insurance agents, and even the guys who paint the lines in parking lots. They all represent the gasoline technology suite.

Similar technology suites exist for trains, aviation, trucking, shipping, and chemicals. Even something as simple as propane has its own technology suite consisting of manufacture, delivery, and tanks for handling. All so that the propane, grill, heater, stove, and lamp manufacturers have the fuel so that consumers can use their products.

What is in that simple barrel of oil? A mixture of compounds, each of which has spawned a technology suite representing millions of jobs in thousands of niche products. Like the box of chip bags, the barrel of oil provides for all the industries, or it provides for none.

Know Your Crude

We hear it on the radio every day, the price of oil, or more precisely price of a barrel of oil. Those outside the petroleum industry track the price, realizing that the cost of gasoline rises and falls based on that "price of oil." But what are they really talking about when they quote the price?

To be honest, there is not "a price of oil" as you hear on the news; instead, there are many prices of oil, each depending on a specific source of the oil being sold. The most common price quotes you hear on American news are the West Texas Intermediate (WTI), the North Sea Brent (Brent), or the OPEC Basket (OPEC). These represent the price of oil produced in various parts of the world (American, North Sea, Middle East) and the quality of the product that comes from those places. There are many prices within the industry itself as each locality produces slight variations in crude, so Gulf of Mexico oil (MARS) is priced differently than the nearby Texas oil. The website oilprice.com tracks the prices of 150 different crude oil blends and indexes, in case you need the specific price of a specific oil. Still, the price of WTI is usually enough to track. But what exactly are all these different types of crude?

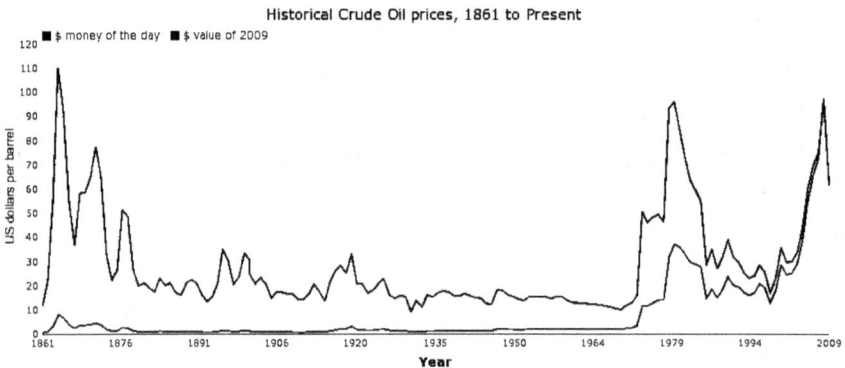

Figure 20 Historical Price of Oil

Essentially, crude oil is a chemical so incredibly complex that it is impossible to determine or express its content utilizing molecular composition or chemical analysis. What is done is to measure certain physical characteristics and infer the chemical composition from that information. Thus, crude oil is evaluated upon two essential criteria: sulfur content and density or viscosity (measured as specific gravity). These two values affect quality and cleanliness.

Table 6 Daily Oil Prices

OIL AND NATURAL GAS PRICES AS OF 11:10 AM CT 05/18/2021

	Price	Change	%Change	Contract
WTI	64.41	-1.86	-2.81%	JUN 2021
Brent	67.56	-0.190	-0.274%	JUL 2021
Natural Gas (Nymex)	3.019	-0.090	-2.89%	JUN 2021

U.S. RIG COUNT FOR MAY 14, 2021:

	Oil	Gas	Total	Year ago Total 05/15/2020
Rig Count	352	100	453	339
Change from previous week	+8	-3	+5	----

*There is 1 rig classified as "miscellaneous"

Whether a crude oil is "sweet" or "sour" is determined by its sulfur content. Sulfur content of 0.5% or greater is called "sour."[108] Legend has it that early prospectors could smell and taste the oil's sulfur level. While oil naturally has a somewhat "sweet" taste, the increase in sulfur will bring the flavor onto the sour side. The EPA, and other environmental regulatory agencies, require that refined products not exceed a set value of sulfur. This requirement means that sour products need to have the sulfur removed before product can be shipped. Clearly, sweeter oils are more likely to be sold into a market with more restrictions, such as the western world, and sour oils into one with less restrictive requirements.

Judging the weight (light, intermediate, heavy) of an oil depends mainly on its API gravity, a measure from the American Petroleum Institute. As the value is based on water, it would be intuitive that "light" oil will float upon the water, while a "heavy" will sink to the bottom. Lighter oils have a higher percentage of hydrocarbons and are easier to refine, while heavier oils require extra processing before running through a traditional refinery process. Since crude oil is a soup of compounds, the specific gravity for a specific crude oil represents the compounds' average weight within that soup.

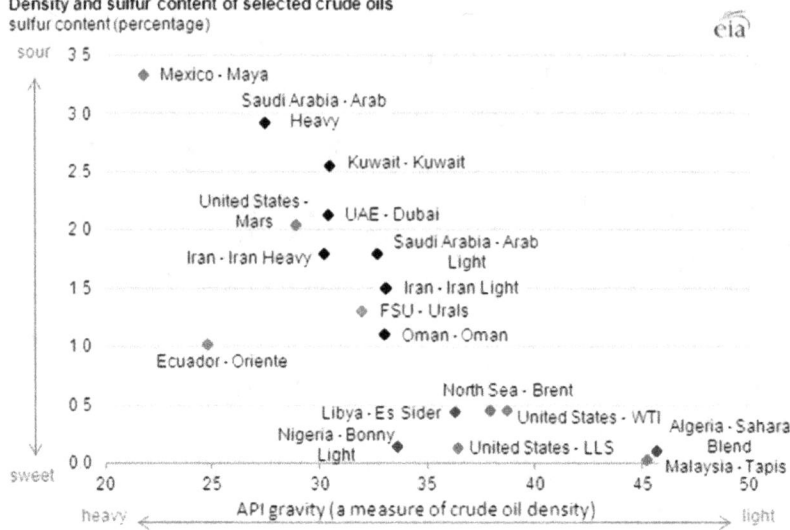

Figure 21 The Array of Crude Oils

Depending on where the source fits on the chart between Sulfur content and Density, it affects the barrel's value and cost. As the barrel contents are affected by the density, different product ratios and the differing value of those products lead to the variable costs. The purchase of different crude types can result in different products, very much as demonstrated by the potato chip box example earlier, as the Green Onion chip lover is more likely to purchase the box that contains more of that flavor.

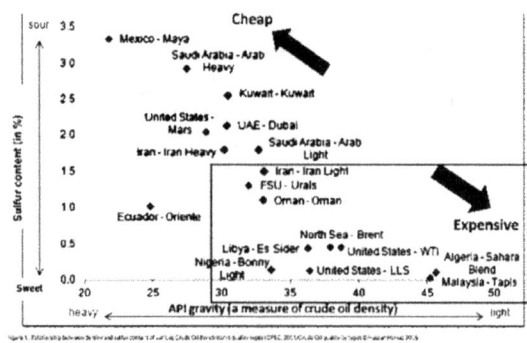

Figure 22 How the Price Varies on Crude

Fossil Fuel Transportation

As a rule, the desired crude oil source is never in the place where it is needed for the next step in the process. Crude oil transportation represents a technology suite that enables refineries to access the best crude oil source depending on the regulatory environment and potential sales of its output.

The required transportation of the raw material utilizes four basic transport methods: ship, truck, train, and pipeline. The use of trucks for over-the-road transportation is limited due to the volume and costs associated with that method. A standard tanker trailer can hold the equivalent of 190 barrels of oil, which means that as an example, when the Prudhoe Bay oil field peaked in 1989 at 2 million barrels a day, the Alaska highway would have needed to support 10,500 tanker trucks heading south every day. When it comes to trains, a 100-car train of tankers can carry about 70,000 barrels of oil (700 per tanker car) at the cost of around $10 to $15 per barrel. Prudhoe Bay would have still required 28 of such trains to move south each day. Thus, the Alaska Oil pipeline was justified as it carries the 2 million barrels cheaper ($5 per barrel) and safer. Of crude oil and petroleum products, 70% are shipped by pipeline, 23% by tanker ships or barges, 4% by truck and 3% by rail.

The right crude to the right refinery with the right transportation mode produces the right mix of final products to drive the technology suites.

Sine Qua Non

Sine Qua Non: (Latin) Without Which Not[109]

As a commercial enterprise, the petroleum industry is profitable if the combined revenue from the sale of all the products refined from a barrel of oil exceeds the cost of extracting the oil from the ground, the refinery process, and packaging the individual components for sale. Without a profit, the whole enterprise is not worth undertaking.

Famously known as "Drake's folly," Col. Edwin Drake drilled the first commercial oil well, expressly for the primary purpose of collecting and developing a product called "illumination oil." As a result of refining the well's product, an oil was extracted that could be used in lamps, a safer and cleaner material than the alcohol-based liquid commonly used. That oil product, which we now know as kerosene, could easily be distilled from the material, and sold at a price much cheaper than the existing burning oils, due primarily to the difference in taxation rates. Although the alcohol was being used as a burning product, it was subject to a vice tax, just the same as the alcohol intended for drinking.

The sine qua non, the reason without which not, of prospecting for oil through the civil war and afterward, was to collect that illuminating oil. Kerosene had gone on to win the battle of the marketplace over whale oil and alcohol-based lamps. The profits from kerosene sales drove the industry to drill for and process more and more petroleum products. Well, at least the profits were great until sales slumped due to the new-fangled thing called the electric light.

As the market for illuminating oils grew, the amount of waste product became overwhelming. In a 42-gallon barrel of crude, roughly 4 to 6 gallons[110] are usable for the illuminating lamps, leaving a vast majority of the material relegated to being disposed of as waste. For the first few decades, with no use for gasoline, it was typically burned at the refinery or simply dumped into the rivers.[111]

Based on the combustion engine developed by Nicolaus Otto, which had operated on the lighter gases, Gottlieb Daimler and Wilhelm Maybach invented a carburetor system[112] that could take the waste product called gasoline to create an aerosol that would burn within the combustion chamber of an engine. When combined with Karl Benz's[113] automotive chassis, the first practical automobile was born.

The automobile craze has never diminished in the past 100 years as Americans and Europeans found new places to explore and new lifestyles to live. There were many miles to be driven, suburbs to be populated, and a desire to get away from mass transit systems and be self-sustaining in transportation. The electric light obsoleted the demand for illuminating lamps. Thus the fuels for the automobile soon became the new sine qua non of oil refining, the thing without which the oil industry would no longer find it worthwhile to continue exploration and refinement.

As more and more oil was refined with gasoline as the intended purpose, new uses for the remaining byproducts needed to be found. Diesel and Aviation fuels soon found homes, and then in the years after the Second World War, technological research saw the development of plastics and petrochemicals. In the 21st Century, America has entered a new phase where petrochemicals are in such demand that they may soon become a new Sine Qua Non for the oil industry.

Imagine two pictures in your mind, one of modern America and one of a third-world country. What are the differences? Largely the difference will be all those things that the developed nations have created from these petroleum-derived products. Those chemicals that were once waste from the refinery process are now the essentials of life in the developed world. With time, scientists and chemists have learned the lessons that reside in a barrel of crude and have brought forth a whole new world.

6

Better Living with Petrochemicals

> *"Interestingly, the primary economic reasons that oil refineries even exist for societies lifestyles and economies are NOT to manufacture the aviation, diesel, and gasoline fuels for today's military and transportation industries. From one forty-two-gallon barrel of oil, only about half is for fuels while the rest is used to manufacture the chemicals derivatives and byproducts that are part of our daily lifestyles."*[114]
>
> Stein and Royal: **JUST GREEN ELECTRICITY: HELPING CITIZENS UNDERSTAND A WORLD WITHOUT FOSSIL FUELS**

In 2016 and 2017, an organization, called the Break Free Movement, organized mass bicycle rides across Europe to stand against fossil fuels and push to accelerate a shift to clean, renewable energy. The object was to demonstrate that we could all be free from fossil fuels; after all, we could all just ride bicycles.[115] [116]

Figure 23 Biking to End Fossil Fuels

But just how fossil-free were they? The modern bicycle frame is usually constructed from tubes consisting of one of several metals, most of which use fuels in their creation (aluminum, steel, titanium, carbon fiber) and either weld or bond the tubes together (adhesives and welding gases both are petrochemicals). Even the wooden and bamboo frames, promoted as "earth-friendly," end up using other materials to make the frame the right size and strength. Now, add to the frame, a pair of wheels made from synthetic rubber with synthetic rubber inner tubing. It is completed with brakes, made from a different synthetic rubber compound, controlled by cables with a plastic coating. Topped off with a saddle made either from leather conditioned with dyes and softeners, or artificial fibers. The stainless-steel drive train runs smoothly with lubricants and greases. Literally, every part on the bike used either the heat from fuel or a petrochemical in its construction. Their personal attire was no different. Bodies clothed by Lycra, spandex, and polyester while heads were protected by helmets made from polystyrene foam covered with plastic outer shells. They managed to top it all off with a large vinyl banner to express their feelings.

That parade in the name of freeing the world from fossil fuels was, in reality, a parade of the wonders of petrochemicals. The riders of bicycles before the development of modern technologies had a name for their machines: the boneshaker.[117] A bike that ran on iron wheels, with blocks of wood for brakes and "small lubrication tanks that would wick oil from soaked lamb's wool into the bearings to help them run smoothly." The riders' backsides from those organized mass ride events must have appreciated the value of petroleum-derived chemicals after all.

As the Break Free movement's bike riders should have realized that it is nearly impossible to avoid being involved with fossil fuels in some way or another. Even those who attempt to live an entirely natural life find it difficult. Birkenstock shoes are popular among that crowd, but they are probably unaware that the leather is treated with chemicals and the soles are actually made from EVA (ethyl vinyl acetate).[118]

The Chemical Revolution

The American Fuel and Petroleum Manufacturers Association (AFPM) describe the petrochemical business as:

> *"If you can imagine it, our products are part of it. Petrochemicals are the building blocks that are essential to making the goods that make modern life possible.*
>
> *All of these things start with just six basic petrochemicals — ethylene, propylene, butylene, benzene, toluene, xylenes — that are combined with other chemicals and transformed into other materials that make products better."*[119]

While Kier and Drake were drilling and collecting oil for that Sine Qua Non of illuminating oil to burn in oil lamps, others sought "city gas," a gas compound routed by pipeline to provide gaslighting for private and public spaces. The gas was obtained by cooking the coal under pressure, for which the resulting waste byproduct was coal tar. As James Burke tells the tale in his book, **CIRCLES**, "Coal tar was a by-product of the manufacture of coal gas, given off while cooking coal. In those carefree pre-ecology days, the tar was happily disposed of by the ton in quarries, rivers, and ponds."[120] Like the waste gasoline, as we saw in the previous chapter, scientists used this basically free waste material for research. With the advent of electricity, this research becomes the core Sine Qua Non of using coal for chemical development.

In 1856, researchers at the British Royal College of Chemistry determined that the coal tar was largely made up of chemicals with a common chemical base, aniline. From this aniline, "some of the first of such products were a whole range of artificial colorants known as aniline dyes, which were then successfully marketed by Germans like Friedrich Bayer."[121] Whose pharmaceutical company we will meet several times later. These aniline compounds could also be added to rubber to create better and longer-lasting tires, which will be important later as well.

A Scotsman, Charles Macintosh, began his own experiments with coal tar, leading to a process to extract naphtha from it. When combining the extracted naphtha with rubber, a mixture was created that we spread over cotton created a waterproof material, useful for raincoats (the British mackintosh) and waterproof tenting. This processing line gave way to the release of olefins, a base chemical that would go on to be used in all sorts of medicine and food additives.

In the 21st Century, these same basic chemical and polymers are being extracted to be used as feedstock, the precursor material for making other things. The techniques have moved beyond coal, so today, oil, natural gas liquids, methanol, and naphtha have all been added to coal as a source of these chemicals.

As you may remember, at the beginning of the lockdowns, the media was quick to report stories noting a shortage of medicines coming out of India because they were not receiving the basic elements out of China. At the moment, China is the largest manufacturer of basic chemicals and polymers, largely made from coal to olefins and coal-derived methanol to olefins processing, creating 70 million metric tons of the material last year. While environmentalists are pushing to close or move the coal facilities, DeLome Fair, CEO of Synthesis Energy, comments, "We expect coal-to-chemicals to keep growing, getting cleaner, bigger, and more efficient."

Once a coal derivative, Naphtha is now collected as an output from the oil refinery process, one of those many compounds that, like the potato chips, you whether you want them or not. Currently, Japan, Europe, and the United States have all been in the process of shutting down naphtha to chemicals production plants, the result being, "this trend has put a lot of surplus naphtha on the market that refinery operators don't know what to do with"[122] There is hope that China will move away from brown (dirty) coal to naphtha. This demonstrates the issue when you stop using one of the products from a barrel, problems are created.

As the AFPM hopes that cyclists will realize, "Petrochemicals make progress possible."

Ten Ways in Which Petrochemicals Make Life Better

> *"Black, gooey, greasy oil is the starting material for more than just transportation fuel. It is also the source of dozens of petrochemicals that companies transform into versatile and valued materials for modern life: gleaming paints, tough and moldable plastics, pesticides, and detergents. Industrial processes produce something like beauty out of the ooze."*

Science, Robert F. Service, Sept. 2019

The growth of petrochemicals is anticipated to be from 16% of consumed oil today to 24% by 2040. Growing at a rate of nearly 6% per year, the developed world uses 20 times more plastics and 10 times more fertilizer than the developing nations. Figure 24 shows that petrochemicals represent the largest expected growth in oil demand in the years until 2030 as it grows by nearly 3.2 million barrels per day.

There are many ways in which these chemicals have made our lives better. Every aspect of life has been affected, and narrowing it down is a challenge, but here is a look at ten areas of our lives where fossil fuels have changed everything. If we consider what represents modern life today, it will primarily reflect what we have made with the flexible petrochemicals.

Better Living with Petrochemicals

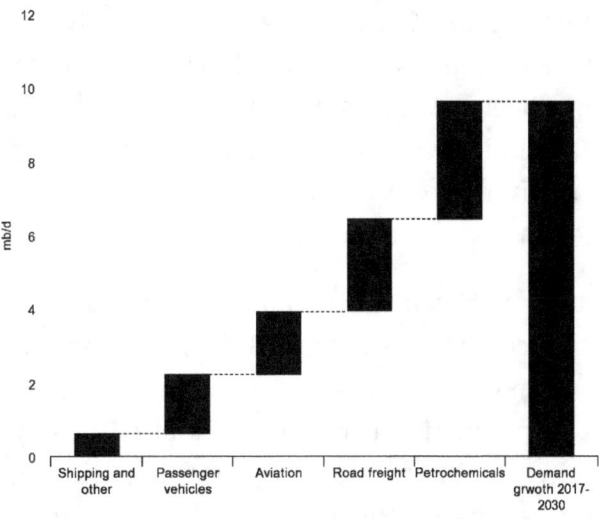

Figure 24 Anticipated Oil Demand Growth by Sector 2017-2030

#1 Giving Strength to Structures

Two of New York City's iconic symbols epitomize the city's transition into the modern industrialized age: the Flatiron Building and the Brooklyn Bridge. The forerunners of future building booms which would soon follow, opening new prospects and industries throughout the nation and the globe. Neither building project could have been undertaken without the two primary materials used in construction: concrete and steel.

Those concrete and steel elements provide the backbones for every structure of significant size, giving them the integral strength to hold the projects together. Originally, New York City required masonry to be used to construct buildings for its fireproof nature. Building codes changed at the beginning of the 20th century and thus was born the Flatiron building, constructed of a steel skeleton – with steel from the oil and coal-powered city of Pittsburgh, PA. Due to the shape and placement of the Flatiron Building, it was subject to accelerated winds, unlike the other buildings of similar construction methods. The building's stability with the relatively lightweight frame convinced the developers of building codes and property developers of the safety of steel and glass skyscrapers. Towers that could never have been constructed except for the strength of the inner core of the concrete and steel elevator shafts around which they are constructed.

"Making steel and cement accounts for around 10% of all emissions,"[123] according to a recent Bill Gates interview, justifying his own investments in changes to how the material is created. Globally, steel production consumes 13% of coal production and emits around 7% of all emissions.[124] Currently, coking coal is used to achieve the heat necessary for the volume of steel required. Small units can create "green" steel but at a 20-30% cost over that from using the current process.

Concrete is considered the second most consumed resource after water. Ironically, both are main ingredients in concrete. Most Americans think of the terms concrete and cement to mean the same thing, but actually, cement is the ingredient that holds the concrete materials together. The beauty of concrete is that "it's affordable, you can produce it almost anywhere, and it has all the right structural qualities that you want to build with for a durable building or for infrastructure."[125] It was the discovery of cement, based on volcanic ash, upon which the Roman Empire was built, enabling those roads and aqueducts that still exist to this day. Cement is created by baking limestone and clay in 1400°C natural gas fired rotating kilns before being ground to a powder by fuel driven machinery.

While concrete has widespread usage, traditional wooden building methods are still dominant outside of commercial and industrial construction. The modern wood-centric building methods depend largely on various wood composites, from plywood to laminated flooring. Although the majority of the material may be wood, it is all held together with adhesives and epoxies. The polymer bases for nearly all adhesives and epoxies available today begin life as isoprene (2-methyl-1,3-butadiene), and while it could be produced from plants, especially rubber trees, it is usually derived from petroleum to achieve the quantities required. Further, the wood is chemically treated to create lasting structures that can withstand weather and pests, yet more petrochemicals.[126]

While major construction projects are fortified by concrete and steel, heavily dependent on fossil fuels, our personal spaces in our homes and offices are subject to many of the same petrochemicals.

#2 Constructing Personal Spaces

Next time you are shopping in your local box store, take a journey into the paint department. While standing in the aisle, take a whiff of the smell. What is that? It is volatile organic compounds (VOCs) that are emitted by paints and varnishes. They include toluene, xylene, ethyl acetate, and formaldehyde.[127] In your home or office, all the colors, from the paint on the walls to the color in the artwork, are the result of dyes that are typically fossil fuel derived, first from coal tar and later petroleum.

Likewise, bring new furniture into your home. Unless you especially seek it out and are willing to pay extra for it, real natural wood furniture is rare. Primarily wooden furniture is actually some form of wood composite or plastic. Unless you allow the items to "air out," you can sense the outgassing of the same chemicals used in the glues that hold everything together. The fabric on the furniture, as well as most window treatments, are constructed of synthetic fibers (see #5: Driving Fashion) and colored by petrochemical-derived dyes.

There are several flooring options within your personal space, and mostly it is a choice between which version of petrochemicals you would prefer. Carpet and area rugs represent more than 50% of all flooring, with 11 billion square feet sold each year, of which less than 5% is made from recycled materials. From the carpet's nylon fibers, stain repellent, antimicrobial, and flame-retardant coatings, to the polyvinyl chloride backing, petrochemicals make up 90% of the bulk of a carpet. The largest market for an alternative to carpet is vinyl flooring, followed by wood laminates, both enabled by fossil fuels.

From plastic plates and glasses to plastic storage bags and containers, the kitchens of the developed nations have cabinets full of these wonder products. Food brought home from the store in plastic bags, whether of the disposal or reusable variety, are packaged in various plastic containers, plastic bags, and plastic-lined cans. The modern containers keep food fresher, safe from tampering and contamination, and shelf stable. In the previous century, food was preserved in a wooden icebox cooled with a block of ice and a fan. Luckily, today's fridges and freezers are nearly maintenance-free due to their plastic construction, blown insulation material, and a cooling mechanism based on the expansion and condensation of freon or similar competitive material. The homeowner is now freed from that frequent delivery of ice, and the food can be kept on the shelf or frozen, sometimes for years.

The home and office are primarily constructed with products that use petrochemicals in their construction, but how did they get from the manufacturer to your home?

#3 Making Transportation Possible

From automobiles to airplanes to bullet trains, each requires fossil fuels for construction and translocation.

For a century, automobiles were constructed from steel and glass. Now, the modern car needs to be fuel-efficient, so they are built lightly with plastics, mostly polypropylene and polycarbonate.[128] The car's interior is all plastic from the dashboard, steering wheel, gear shifter, floor mats, ceiling/door coverings, carpet, and polyester seats.

There is one thing that nearly all of the 2.2 billion road vehicles in the world have in common, they roll on tires made from synthetic polymers and synthetic fibers. According to the U.S. Tire Manufacturer Association website:[129]

> *The two main synthetic rubber polymers used in tire manufacturing are butadiene rubber and styrene butadiene rubber. These rubber polymers are used in combination with natural rubber. Physical and chemical properties of these rubber polymers determine the performance of each component in the tire as well as the overall tire performance (rolling resistance, wear, and traction).*
>
> *Another important synthetic rubber is halogenated polyisobutylene rubber (XIIR), commonly known as halobutyl rubber. This material causes the inner liner to be impermeable, which helps to keep the tire inflated.*

Further, the 4% of the tire labeled in figure 25 as textile "are polyester cord fabrics, rayon cord fabric, nylon cord fabric, and aramid cord fabric." All of which are man-made fibers. The 26% filler consists of carbon black and precipitated silica which provides the wear and abrasion characteristics of the tire. Originally, rubber tires were white as rubber is naturally white, but additives make them both longer lasting and black. Such Carbon black is created by the "reaction of a hydrocarbon fuel such as oil or gas with a limited supply of combustion air at temperatures of 1320 to 1540°C (2400 to 2800°F)." In other words, the purposeful creation of soot for use in many products, including car tires, inks, paint, and dyes. So it turns out that your black ballpoint pen and your car tire, both made by burning fossil fuels.

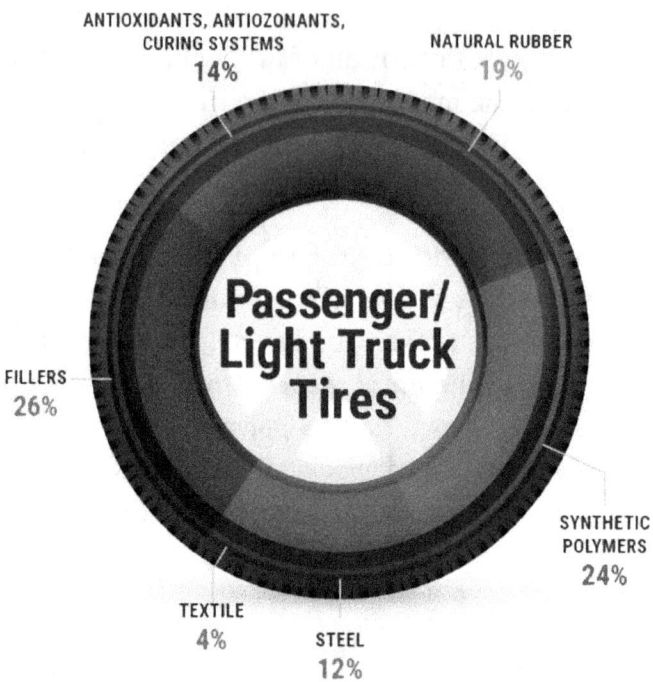

Figure 25 Tire Composition

When it comes to flight, weight is everything. Every pound of weight in the aircraft is a pound of paying passenger or cargo that cannot be carried. As Samoa Air CEO Chris Langton said, "What makes airplanes work is weight. We are not selling seats; we are selling weight." A Boeing 757 weighs 62 tons empty and has a maximum takeoff weight of 126 tons. Virgin Atlantic estimated that reducing a single pound of weight could save 15,000 gallons of fuel each year. [130] Weight savings on the order of tons are possible by using plastics, including the seats, overhead compartments, the walls and ceiling, and floor coverings. Those windows that give you a view of the city lights upon landing are layers of plexiglass with plastic covers.[131] When settling into your next in-flight meal, consider when you lower your plastic tray table to eat from a plastic plate with plastic utensils, just how much plastics have made your flight possible.

Much of Europe and Asia utilize bullet trains for their high-speed transport, and while not as sensitive to weight as flight, the ability to maintain those speeds does mean that weight must be minimized, and that again is done with interior construction with plastics. A Materials Today article entitled, "Can Trains be half plastic?" notes, "This is especially so for interiors where they are used for items as various as side panels and trim, seats, tables, window surrounds, baggage racks and bins, bulkheads, standbacks, floors and ceilings, vestibules, toilet compartments, and staircases. Cumulatively, they can save considerable weight, yielding worthwhile performance and economic benefits."[132] The result is that "A 10% saving in the mass of a metropolitan rail vehicle can reduce energy consumption by 7%, saving up to $100,000 annually per vehicle." So that the bullet trains can reach their high speeds safely, quality steel rails are required, whose strength comes from the use of coal.

Even the junk contents with which we litter our cars, planes, and trains are likely also derived from petrochemicals, including Styrofoam burger boxes, coffee cups, plastic soda bottles, water jugs, shopping bags, candy wrappers, deflated sacks of chips, Saran Wrap, sandwich baggies, Igloo thermoses, melted ice packs, beach chairs, and foam pads for sleeping bags.

As the design and construction of autos have converged to a common, market-driven solution, one of the main selling points used to differentiate the products has become infotainment. The merger of information, from car performance to traffic alerts and entertainment, from radio and CDs to movies and mobile phone interfaces. The automobile has become a mobile version of our technology obsession.

#4 Technology Age

This very book could not have been written without the benefit of fossil fuels. True, it could have been written on a typewriter from the 1880s, but that wasn't going to happen. Even then, the typeface would likely have been made from steel enhanced by coal.

Addicted to devices and entertainment, the social network of the globe revolves around electronic devices. Those devices depend on semiconductor electronics which in turn depend on fossil fuels for manufacture and use. One study on the environmental impact of semiconductor devices indicated that the total weight of secondary fossil fuels required to manufacture a 2-gram integrated circuit was 1.7 kilograms.[133] Taiwan, the home of Taiwan Semiconductor Manufacturing Company (TSMC), the world's largest semiconductor manufacturer, reports that it uses 17.7% of its electrical generation in manufacturing semiconductor devices.

So that silicon semiconductor enabled gadgets can travel with us, in nearly any environment, they must be protected from corrosion which is accomplished through a molding process. This process consists of using "epoxy resins, phenolic hardeners, silicas, catalysts, pigments, and mold release agents."[134] Much of which begins life in the crude oil soup. As a result of the miniaturization of electronics, the molding compound has been replaced by a conformal coating, but that is also a petroleum derived polymer.

Nearly all of our larger entertainment devices, from computers to televisions, use these integrated circuits mounted onto fiberglass circuit boards held in plastic shells. From plastic coated wiring to plastic screen protectors, our entertainment is brought to us through fossil fuels.

Our lives now depend on having our technology devices always at hand. They have become part of the fashion of our lives. Plastic covers with designs and decorative features, often carefully selected to make a statement, have become part of our attire.

#5 Driving Fashion

In Hamlet, Polonius tells his son Laertes, "The apparel oft proclaims the man!" Or in the more modern version, "Clothes make the man." From the ancient time of Homer up until today, the influence of fashion has driven man to seek out the latest in clothing. In many ventures, the ability to be in the proper attire at the appropriate time can be the difference between success and failure. From CEOs to politicians, the thousand-dollar suit has become one of legend. One of the more popular exhibits at the Smithsonian is the collection of inaugural ball gowns worn by the first ladies. Why? Because humanity has always been obsessed with the fashion of the rich and famous and, where possible, try to emulate it. When Melania Trump wore a $2,190 Roksanda dress to the 2016 Republican National Convention, it sold out less than an hour after the speech.[135]

The $1.5 trillion clothing, apparel and footwear industry is so important that from production to sales, it employs one out of every six people worldwide. Upon the fashion market's strength, countries such as Vietnam, Bangladesh, and Sri Lanka have become developing countries. Textiles represent 20% of Bangladesh's GDP and a full 80% of the country's exports. Without these exports, these countries' economies would be in danger as Siddiqur Rahman of the Bangladesh Garment Manufacturers and Exporters Association said, "Our economy is dependent on it."

Clothing is a basic human need for survival, except for those few communities living near the equator, which National Geographic likes to highlight. Survival during the northern hemisphere's winters throughout most of human history has hung upon the access to clothing, which was generally made of just a few materials: skins, furs, silk, cotton, wool, and linen. Before the petrochemical industry, only natural materials were available and usually limited to those locally accessible. Among the hunter-gathers of the past, the hunters collected the skins and furs they could catch. At the same time, the gatherers created fabrics based on agriculturally based materials. As such, linen and cotton were widely available in those areas where its cultivation was possible, requiring large tillage areas and a suitable climate of plenty of sun and being relatively frost-free. Meanwhile, silk production posed a more difficult proposition as it involves access to the silkworms and the necessary mulberry plants.

Roughly 2.5% of all croplands is used to raise the 25 million tons of cotton production, of which 63% is used in clothing, 29% in home furnishings, and 8% in industrial products.[136] The projected use of cotton in clothing is expected to remain flat in the future due to growth in the synthetic fiber markets.

The first synthetic fiber development began in the late 1890s when cellulose fibers were derived from wood and plants to create artificial silk and later rayon. Today fibers synthesized from naturally occurring cellulose represent 6% of the market. The major sources for creating the cellulose base mixture are wood pulp, primarily pine, spruce, or hemlock trees.

Petrochemical-based clothing fibers came onto the market from several directions, with the appearance of nylon in 1931, Dacron in 1946, polyester in 1951, and spandex in 1959. All of these fibers are derived from a combination of coal, water, and oil. These fibers are sturdy, reliable, presentable, and cheap. Today, these synthetic fibers represent 52% of all fibers used in fabric creation.[137] In 2014, U.S. imports of synthetic fabric overtook imports of cotton. While cotton fiber growth is flat, synthetic fibers are growing at 8% per year.[138]

The clothing industry has moved solidly away from natural fabrics in favor of synthetic ones, with 72% of clothing produced in 2019 based on synthetic fibers. The result is that the fashion industry now consumes 25% of the industrial petrochemicals for its fibers, dyes, and coatings, resulting in nearly 10% of all carbon emissions.

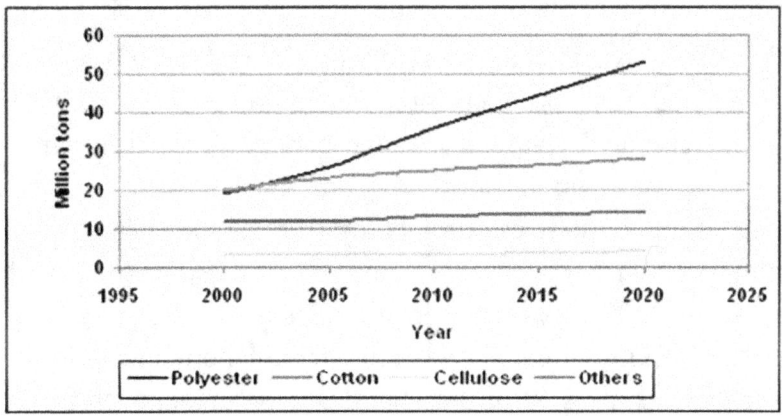

Figure 26 Fiber Market Growth

It is estimated that more than 100 billion individual items of clothing, primarily mass-produced from synthetic fibers, are produced each year, with as much as 20% going unsold.[139] The unsold items, so-called "deadstock," are typically incinerated or defaced. The huge volume of these discards represents 62 million tons of product. Typically, of the 80 billion items sold, each will be worn an average of only seven times before being relegated to the trash bin.

Ironically, Americans spend $1700 per year to have far more clothes than they would ever need. A recent survey indicated that Americans believed that they had used 56% of their wardrobe in any given year when in actuality, they used just 12%. In 2015, the average woman had 30 outfits in her closet, compared to only 9 for her 1930 counterpart.[140] When Martha Jefferson, the wife of the future President, passed away, an inventory was done of her wardrobe, which consisted of 16 dress gowns and 4 daily work gowns. The average American woman today has twice the outfits of the first lady of Virginia two centuries ago. According to the EPA, Americans dispose of 10.5 million tons (70 pounds per person) of fabric materials each year, most of which contain non-biodegradable synthetic fibers, 95% of which could have been recycled,[141] assuming there was an industry to handle it all.

Consumers have flocked to synthetic fibers for their moisture-wicking and easy product care. They tend to be wrinkle-free, stretch-free, and generally long-lasting. The difference in appearance between the start of life and the end of life can be quite small, as opposed to natural cotton, where material can show significant wear with use. Imagine the collapse of the torn jean market if everything wore like synthetic fibers. The average mother spends 55 minutes per week ironing those non-synthetic fibers, which results in a strong demand for permanent press. From swimming attire to yoga pants, synthetic fabric has enabled fashion that natural fabrics can never emulate.

Rabbit Clothing, a boutique online clothing manufacturer, recently introduced their quick n' fit ICE line of athletic clothing touted as sustainable and environmentally friendly. A brief review of their website reveals the composition of the material: 38% recycled nylon, 19% virgin nylon, 29% polyester, and 14% spandex. So essentially, their eco-clothing is still made from petrochemicals, even if 38% has been recycled. Now, you may read that they use recycled nylon and think about collecting some of those 70 pounds of disposed of clothing that has been worn and recycling it, but this would be incorrect. The recycled nylon producers gather incomplete, unassembled, or reject pieces from the stitching houses, along with some of the 20% of items that never sell and recycle those clean pieces. This doesn't remove used clothing from the refuse stream but removes unused cloth. The cost of sorting and cleaning used fibers is too significant to justify its use. This demonstrates that regardless of the advertising, getting away from synthetic fibers is difficult.

There is a movement to move away from synthetic fibers and back to cotton, particularly organic cotton. Currently, organic cotton represents just 6% of clothing sales in the U.S.. The production of organic cotton is limited to only 15,000 tons (0.06% of the total). Clearly, from the numbers, clothing is made with organic cotton and not of organic cotton cloth. For this movement to be anything other than a small niche market would require significant increases in land usage for growing cotton, on the order of an additional 7.5% of cropland. While considered natural, cotton takes a toll on basic water requirements when considering the entire process from growing through processing; approximately 1000 gallons of water are required to produce a single cotton t-shirt.

The elimination of fossil fuels would result in significantly fewer clothing items; thus much higher prices, limited choice, and increased land and water use to produce those items. All the while, it would destroy the growing economies of developing countries and all those who depend on the fashion industry. But the selection of fabrics isn't always just about fashion, often there are practical matters to consider.

#6 Exploring the world and universe around us

All these new fibers had the added benefit of opening the world of exploration to the everyday person. The first men to reach the top of Mount Everest were Edmund Hillary and Tenzing Norgay in 1953, unless Mallory and Irvine made it to the top in 1924, a descent of which they did not survive to tell us. But with the advancement of materials from tents to clothing, all made from synthetic fibers, approximately 300 climbers currently reach the summit each year. All of those 300 climbers are equipped with items manufactured from some version of Gore-Tex, "a durable, breathable waterproof and windproof fabric ... made from expanded polytetrafluoroethylene (ePTFE for short)." From the synthetic rubber compounds in the treads of their boots to the foam and plastic of their safety helmets, the mountain climbers, much like the protest cyclists, are clothed from top to bottom in materials dependent on fossil fuels.

Outdoorsmen and mountain climbers everywhere are dependent on just a few manufacturers of outdoor gear. One of the largest is the VF corporation subsidiary, The North Face, producer of award-winning clothing, jackets, and tents. Its pricey apparel is largely made from petroleum-based fibers, shipped by fossil fuels, and displayed on sales racks largely constructed of plastics. Still, it recently decided to add the oil and gas industry, to a list of industries to which it would no longer sell due to "brand standards," a list that previously included guns, tobacco, and porn. As such, it famously rejected an order for embroidered jackets from Innovex Downhole Solutions, a West Texas manufacturer of equipment for oil and gas drillers. While North Face is busy virtue signaling against the oil and gas industry, the Colorado Oil and Gas Association mockingly awarded North Face with their customer of the year award for their innovative use of fossil fuel derivatives. It is a fine line to walk selling to those who love the outdoors while promoting an eco-friendly image, while all the while benefiting from the wonders of petrochemicals.

The same advanced material has been used to weave the spacesuits used for all the explorers of space, from the fabric to the hoses that deliver oxygen and cooling to the helmet and face shield. All of which would not be possible without the availability of fossil fuel-based materials. The materials to construct an Apollo suit is described as:

> *"Numerous raw materials are used for constructing a spacesuit. Fabric materials include a variety of different synthetic polymers. The innermost layer is made up of a Nylon tricot material. Another layer is composed of spandex, an elastic wearable polymer. There is also a layer of urethane-coated nylon, which is involved in pressurization. Dacron—a type of polyester—is used for a pressure-restraining layer. Other synthetic fabrics used include Neoprene that is a type of sponge rubber, aluminized Mylar, Gore-Tex, Kevlar, and Nomex."*[142]

Of course, the explorers need a craft in which to ride during their journey into space. The vessel in which those explorers ride is also only possible through materials based on petrochemicals. A common item used inside spacecraft, ever since its development for that purpose, is Velcro, made from nylon and plastic. Still, the capsule would never get off the ground without fuel. The SpaceX Falcon 9 booster currently uses liquid oxygen and rocket-grade kerosene (RP-1) to create more than 1.7 million pounds of thrust to launch into space and return with its controlled landings.[143]

One of the promises of space exploration was the possible medical applications that could be available through zero-gravity environments. Many medical imaging techniques have their origins in work done for the space program.

#7 The Medical Miracle

> *"The same U.S. politicians that are thrashing on the oil and gas industry, and seeking its demise, are the same ones reaping the benefits of the medications, medical equipment, communication networks, and the thousands of other products from that industry that have contributed to their lifestyles and their ability to live beyond eighty years of age."*[144]
>
> Stein & Royal, **JUST GREEN ELECTRICITY**

From Samuel Kier's first venture with medicine oil to the latest in advanced materials, petrochemicals have been pushing the health care horizons. By 2020, pharmaceuticals represented 3% of all petrochemical use,[145] and the 3.4 billion pounds of health care plastics represent 4% of all plastics used.[146]

My health is much better from my own personal experience due to these wonder chemicals being used in the medical field. The fix to my own heart attack was to install a medicated arterial stent, using a small balloon to open the blocked coronary artery so that blood could flow again—the stent, it's medicated coating, and the small inflatable balloon all required fossil fuels to be produced. The pacemaker implanted afterward has an MRI-friendly plastic body sealed with epoxy, and the interior electronics are sealed with a conformal coating, all made from petrochemicals. It should be further pointed out that neither of my procedures would have been possible without fluoroscope dyes, IVs, oxygen masks and tubing, tapes, and adhesives. As recommended for all heart patients, I take my 81mg tablet of aspirin every morning to prevent another heart attack. Worldwide an estimated 58 billion tablets of aspirin are taken annually[147] to treat everything from fever to stroke. The medication is based on the chemical salicin, which is found in willow bark. But harvesting enough willow bark for those 58 billion tablets would be impossible, so thankfully, German chemist Felix Hoffman experimented with that waste product, coal tar, to synthesized acetylsalicylic acid in 1897 while working for Bayer Pharmaceuticals. Today, most of the manufacturing of aspirin from organic feedstocks occurs in India and other developing nations. In the end, the health care I received started with transportation of myself, the staff, and all those fossil fuel derived products to a hospital and ended with me writing this book when so many of my male ancestors perished from the same condition.

The ability to lengthen our lives and make more of those lives is primarily thanks to the advances in the world of health care, a field primarily driven by the advances in petrochemicals. As much as 99% of all pharmaceutical feedstocks and reagents begin life in the oil barrel,[148] including essential medications such as "analgesics, antihistamines, antibiotics, antibacterial, rectal suppositories, cough syrups, lubricants, creams, ointments, salves, and many gels,"[149] as well as much of the medical equipment, especially the one-time-use supplies, such as tubing, syringes, splinting and casting. From artificial heart valves to the artificial knees and hips, the devices all use Polytetrafluoroethylene (better known as Teflon). Globally, $1.3 billion is spent annually on Prosthetics, mainly constructed of plastics and carbon fiber material.

A team from the German chemical company IG Farben was working with the same coal tar dyes used in the synthesis of aspirin when they developed the sulfa family of antibiotics in 1937. Treatment with the new antibiotics was responsible for a 25% reduction in maternal mortality, 13% reduction in pneumonia/influenza mortality, and 52% reduction in scarlet fever. Once a major cause of death in children, and when not fatal, it still often left behind kidney and heart tissue damage, but today, scarlet fever is rarely fatal if caught early. The sulfa drugs have largely been supplanted by penicillin, moving from coal tar to petrochemicals. Medics equipped with these two medications were able to improve the survival rate of the wounded to 50% during World War II, as opposed to just a 4% rate during World War I. Further improvements in medications and medical techniques increased survivability to 92% by 2016. Just a century of medical advancement moved from just a 4% chance of making it to just an 8% chance of not surviving.

Non-petrochemical-based products such as gloves, tubing, gowns, and drug delivery devices are being developed, which use renewable raw materials. [150] However, "Cost concerns, regulatory wariness, minimal selection and a lack of strong demand from end-users have deterred medical device manufacturers from pursuing the use of more environmentally friendly plastics."[151] Still, other suggested items, such as implants and hernia repair materials, may have a steeper hill to climb before acceptance happens, as many consumers, while supporting recyclable plastics, may not want them implanted.

One of the leading research fields showing promise is creating body replacement parts either by 3D moldings or printings. Each method requires plastics. The promising field of 3-D printing of a heart depends on polymers called PEEK (polyetheretherketone), which is an implant-safe material that won't be rejected by the body. Recently, a New Jersey man received a skull implant to replace a damaged piece using the PEEK material. Where the implant does not require custom printing, such as heart valves, polypropylene materials are used to build or coat the necessary structure.

It would be impossible to mass-produce pharmaceutical drugs and devices in sufficient quantities to meet the global demand without using synthesized materials. While the Australian Senate recommended a ban on single-use plastics by 2023, they did have one exception: in the use of medical plastics. The typical blood test requires a pair of gloves, a plastic syringe, and one or more plastic vials, all single-use to reduce cross-contamination and infection. The World Health Organization estimates that when glass syringes are used, as many as 30% of the injections are unsafely contaminated. Truly, single-use medical plastics save lives.

Once upon a time, a doctor could carry a collection of medications in his doctor's bag during house calls. Today, the power to synthesize with various feedstocks allows the proliferation of medicines and medical equipment so that the doctor's bag is no longer practical. Now there is a pharmacy with shelves lined with medications and variations, time-released, coated, capsule or tablet, all thanks to the chemicals that come in an oil barrel.

#8 The Food Revolution

> "The battle to feed all of humanity is over. In the 1970s, the world will undergo famines – hundreds of millions of people are going to starve to death in spite of any crash programs embarked upon now."[152]
>
> The Population Bomb, Paul Ehrlich, 1968

Mr. Ehrlich proposed in his famous book that the world would never be able to feed more than 5 billion people. As of 2020, the world supports 7.6 billion people, and although there are still the hungry and starving, the numbers needing food are lower than ever before in history. All thanks in large part to fossil fuels.

These fuels empower the production of food in many ways, from mechanized farming to improved fertilizers. As Alex Epstein says, "Fossil Fuels are the Food of Food."[153] As we have seen before, it took 300 hours and 5 to 10 acres in the agrarian age to grow those 100 bushels of wheat. After the green revolution, not only were the hours required to produce the wheat reduced to under 10 hours, but the land necessary also shrank to 0.5 to 1.5 acres. The current record wheat yield was accomplished by Eric Watson on his New Zealand farm by achieving 258 bushels per acre using special seeds by another division of a company we have seen before, Bayer CropScience. Improved fertilizers, pesticides, and herbicides, all based on fossil fuels and fuel-powered planting and harvesting equipment, all helped American wheat production double since 1961, while the land planted declined by 25%. The average crop yield per acre went from 20 bushels to routinely being over 200 bushels in today's world.

Figure 27 U.S. Wheat Yields 1961 to 2018

With the advent of the new techniques allowing more crops on less land, the cost of food has fallen by up to 75% over the past century. While the United States spends 6.4% of its income on food, these least developed nations still spend upwards of 40%. Throughout history, it was often the best that man could do to earn enough to pay for food and shelter. But life in the 20th and 21st centuries has become unlike any time in previous history, humanity finally had the money for far more than just food and shelter.

With the advent of the gasoline-powered tractor in 1902, it rapidly replaced the draft animal, pushing productivity higher and producing more food than ever before. Tractors peaked at 5.5 million in 1966, and the number has been stable at around 4.5 million in recent years, along with 325,000 combines. Without needing the manual labor to farm or manage the animals, the share of the workforce employed on the farm fell everywhere.

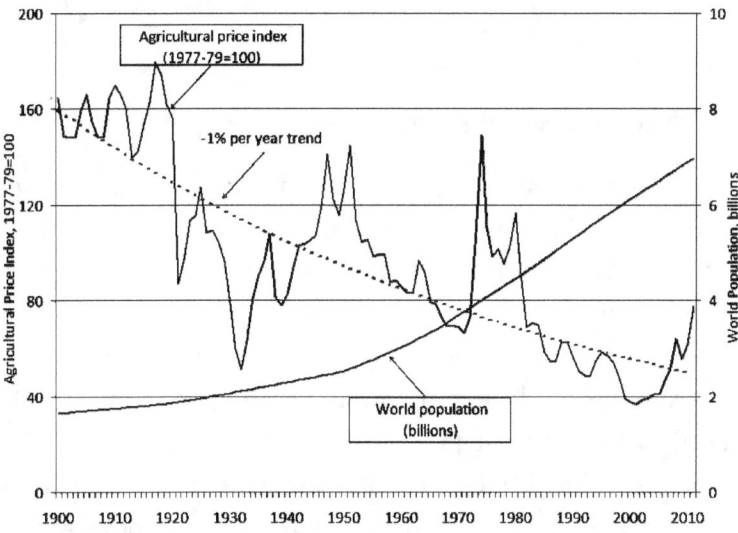

Figure 28 Agricultural Price Index 1900-2010

Any walk through the aisles of the grocery store is really a walk through the benefits of fossil fuels. Coloring, flavoring, and fragrances were all developed from coal tar compounds and mass-produced using fossil fuel feedstocks. A wide array of products made with chemicals to enhance stability and freshness while packaged in plastic containers, bags, and wraps to keep food fresh and uncontaminated. Vegetables and fruits are coated in waxes, often from these compounds. The reason why there is healthy food without purchasing directly from the farmer is those chemicals.

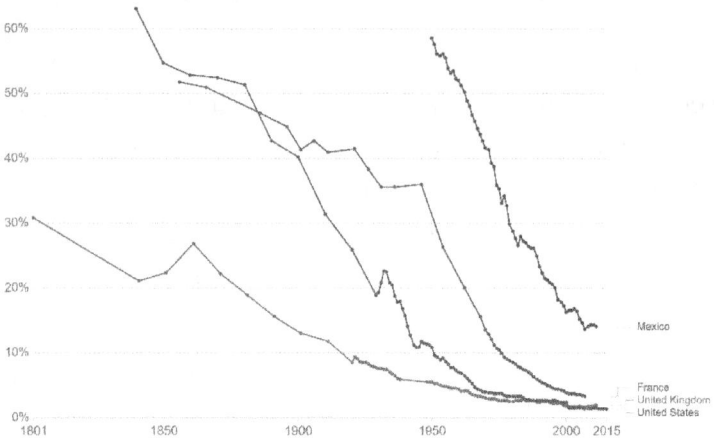

Figure 29 Share of Agriculture in Total Employment

Scanning the ingredient lists of most consumer products these days will include one of many glycol variations: propylene, polyethylene, and butylene. While butylene glycol can be derived from corn-based alcohols, generally, all three glycol products begin life as a petrochemical—these glycols are commonly used in all sorts of foods, cosmetics, and cleaning products. Propylene glycol is found in literally thousands of products, an estimated 75% of all food products, including dried soups, seasoning blends, marinades, salad dressings, baking mixes, soft drinks, flavored teas, powdered drink mixes, and even dog food. The typical uses of propylene glycol in foods are Anti-Caking Agent, Antioxidant, Carrier, dough strengthener, emulsifier, moisture preserver, processing aid, stabilizer and thickener, and a texturizer. The reason why a condiment can sit in a bottle for months and still flow smoothly out is the glycols.

#9 Making Us Beautiful

The first petrochemical was developed in 1872 and soon became a staple of the beauty industry, that being Vaseline. Men and women have used various products to look and smell better in an attempt to be more attractive, look younger, and generally be more tolerable to each other. These products can generally fall into the categories of cosmetics, perfumes, and personal care.

Over a lifetime, the average American woman spends $313 per month ($225,360 lifetime), and the average man spends $244 per month ($175,680 lifetime) on beauty spending.[154] Overall, Americans spent nearly $50 Billion[155] on cosmetic products in 2019, which provides the impetus for the continual creation of products, most of which will contain at least one ingredient derived from petroleum, not counting the plastic used in bottling, transporting, and displaying the products.

Most modern perfumes are derived from plant or animal oils. Still, the process by which these oils are extracted from the original material usually involves dissolving through a solvent (toluene, benzene, or Petroleum ether are commonly used), with the dissolving agent later evaporated away.[156]

Psoriasis suffers everywhere are thankful for the concept of coal tar shampoos. The direct use of coal tar in the shampoos keeps the psoriatic patches under control. However, besides the coal tar on the label, the shampoo also contains those fragrances and coloring that come from the feedstock chemicals.

Those propylene and polyethylene glycols aren't just for food, they are also found in all sorts of beauty products, including deodorant, hand sanitizers, toothpaste, and most creams. I recently noticed the presence of propylene glycol in an "all-natural" body wash product from the Dead Sea, imported from Israel. No matter how remote or natural, those glycols are there to help the material remain fluid at the right density to work well with a pump.

The American Contact Dermatitis Society names an allergen of the year, and over the past 20 years, 15 of the products were petrochemicals, and in 2018, the allergen of the year was propylene glycol which can cause skin reactions and sores to those who are allergic. Other chemicals so named include Acetophenone azine, isobormyl acrylate, and Benzopheneones. The society notices the reactions to these compounds due to their extensive use in beauty products.

#10 A Life of Leisure

A byproduct of society's advancements is that members of the developed communities enjoy life with time and money for leisure activities that our ancestors could never conceive. Since the 1960s, productivity improvements mean that Americans spend 7.9 hours per week less time working and doing chores, which they can spend on leisure activities.

Imagine a trip to the beach. The trip could include a time of surfing on a fiberglass board encased in a polyester resin, made to float through a polyurethane foam core, while wearing a wetsuit constructed of neoprene.[157] You could further take a break by relaxing under a beach umbrella made resistant to UV rays through synthetic fibers, while spreading petrochemical-based sunblock on your skin. All while enjoying a plastic bottle filled with a cold drink from your plastic cooler, cooled by ice which was frozen using refrigerants.

After a long day on the beach, imagine settling into your nylon tent with waterproofing by perfluorocarbon (PFCs), laying back onto your polyester sleeping bag, atop your foam mattress, and settle into a nice sleep courtesy of fossil fuels.

Leisure products constructed from fossil fuel materials include boats, planes, canoes, and kayaks and their paddles, along with helmets and life jackets. Teenagers enjoy molded wheels for skateboards, scooters, and rollerblades. Their helmets, knee, and elbow pads are all plastic. Musicians enjoy woodwind and stringed instruments, piano keys, guitars and drums, and sticks based on fossil fuels. And those who don't play can still enjoy their music through vinyl records, CDs, cassettes, and radios.

Dr. William Roberts sums up the point on just how these fuels have affected our lives, "Since nearly all plastics, polymers, ink, paints, fertilizers and pesticides are made from petrochemicals, and all products are delivered to market by trucks, trains, ships, and airplanes, virtually nothing in our offices or on our bodies is not dependent on fossil fuels."[158]

Every aspect of life which leads man to believe that life is better today is due to fossil fuels, and those material possessions are constructed from chemicals. Dow Chemical once used the slogan, "Better things for better living through chemistry," which is true, especially if they are petrochemicals. There is little in our lives that would remain if the flow of petrochemicals suffered interruption.

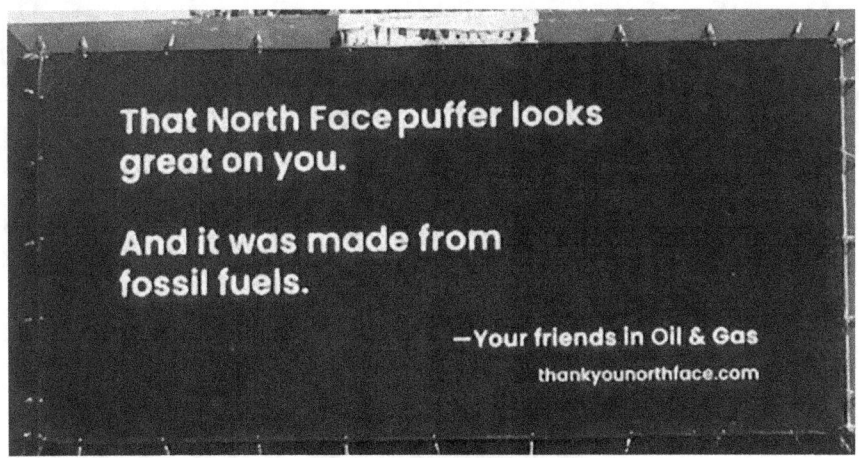

Billboard in Colorado

Many Things Made from Petrochemicals!

Epoxy	Movie Film	Shag Rugs
Deodorant	Bearing Grease	Wire Insulation
Refrigerators	Fishing Boots	Tents
Roofing	Boats	Shoes
Cold Cream	Floor Wax	Skis
Ammonia	Trash Bags	Pajamas
Roofing Shingles	Insecticides	Detergents
Fishing Lures	Fertilizers	Fishing Rods
Diesel	Hand Lotion	Dishes
Candles	Golf Balls	Yarn
Football Helmets	Electric Blankets	Tennis Rackets
Loudspeakers	Artificial Turf	Fan Belts
Vitamin Capsules	Lipstick	Refrigerant
Gasoline	Basketballs	Toothpaste
Roller skates	Bicycle Tires	Synthetic Rubber
Wheels	Sunglasses	Toolboxes
Hair Curlers	Cameras	Paint
Glycerin	Drinking Cups	Antihistamines
Hearing Aids	Balloon	Plastic Wood
Dashboards	Luggage	Dolls
Vaporizers	Garden Hose	Paint Rollers
Shaving Cream	Shower Doors	Dice
Toothbrushes	Ink	Skate Wheels
Antifreeze	Nail Polish	Cortisone
Dyes	Linoleum	Bandages
Slacks	Tires	House Paint
Telephones	Ice Cube Trays	Petroleum Jelly
Glue	Cassette Tapes	Speakers
Ice Buckets	Solvents	Life Jackets
Ice Chests	Soap	Food Preservatives
Paint Brushes	Mops	Dentures
Artificial Limbs	Tool Racks	Dresses
Nylon Rope	Transparent Tape	Insect Repellent

Better Living with Petrochemicals

Rubber Cement	Model Cars	Yoga Pants
Rubbing Alcohol	Shampoo	Plywood Adhesive
Linings	Beach Umbrellas	Ballpoint Pens
Clothesline	Car Battery Cases	Disposable Diapers
Milk Jugs	Salad Bowls	Sweaters
Perfumes	Oil Filters	CD Player
Denture Adhesive	Toys	Awnings
Purses	Safety Glasses	Band-Aids
False Teeth	Safety Glass	Putty
Car Enamel	VCR Tapes	Footballs
Water Pipes	Shoe Polish	TV Cabinets
Heart Valves	Curtains	Toilet Seats
Carpeting	Guitar Strings	Hair Coloring
Credit Cards	Soap Dishes	Percolators
Crayons	Folding Doors	Golf Bags
Upholstery	CD's	Cell Phone Case
Sports Car Bodies	Anesthetics	Laminate Flooring
Motorcycle Helmet	Motor Oil	IV Bags & Tubing
Surfboards	Caulking	Toys
Electrician's Tape	Panty Hose	Sunscreen
Enamel	Combs	Storage Bins
Pillows	Faucet Washers	Car Bumpers
Football Cleats	Umbrellas	Computer cabling
Aspirin	LP Records	Soda Bottling
Clothing Hangers	Clothes	Plexiglass Shields
Parachutes	Dishwasher	N95/KN95 Masks
Antiseptics	Shower Curtains	Hand Sanitizer
Dishwashing Liquids		Computer Keyboard and Mouse
Car Sound Insulation		Eyeglasses and Lenses
Soft Contact Lenses		Food Packaging
Permanent Press Clothes		Foam Meat Trays and wrap
Unbreakable Dishes		Sailboat and Sails
Refrigerator Linings		

7

Fossil Fuel Success = Man's Success

> *"The world as we know it – so prosperous that "poverty" now means someone who is richer in real terms than kings and tsars and popes of yore; so without hunger that millions turn down food every day because it isn't "vegan" or "whole" and so without war that for the past half-century, the only people to have worn a uniform in American are those who have volunteered to do so."*[159]
>
> **THE WOKE SUPREMACY**, Evan Sayet, 2020

As Phil Farrand explained in his book, **THE NITPICKERS GUIDE TO STAR TREK**, when "any culture experiences technological developments in one area, this usually leads to technological developments in other areas."[160] Farrand was always bothered by the depictions on the Star Trek TV show of civilizations that, while they had interstellar warp speed travel technology, in every other way, they seemed stuck in ancient Rome or Greece. Logically, technology should not work that way. When a society makes gains in space travel or other high technology, it will also make gains in other areas like medicine, food, and shelter. Our space programs are an example of this effect. The moonshot of the 1960s drove technology that created all sorts of side benefits beyond just rocket technology, such as freeze-dried foods, space blankets, silver-zinc batteries, and medical diagnostic equipment.[161] The space program further pushed developments in solar panels, pacemaker and defibrillator heart devices, CAT and MRI scanning equipment, and precise timing devices.[162] But it wasn't the first technological windfall that humanity has experienced.

When the industrial age began in the Western Democracies during the middle of the 18th century and subsequently spread around the globe, the world experienced a previous explosion of technology. The development was not limited to just the industrial realm but spread across many fields of expertise, just as Farrand predicts. As the industrial age began to improve lives and change the world, it got a secondary boost with advancements in fossil fuels through the 19th and 20th centuries, which powered yet a further explosion of ideas, technologies, and discoveries. Along with society's industrialization came gains in health/life expectancy and wealth/poverty reduction. As Stein and Royal put it, "The market economy powered by the products from fossil fuels and electricity unleashed wealth, innovation, employment, and freedom unlike any single force in history."[163]

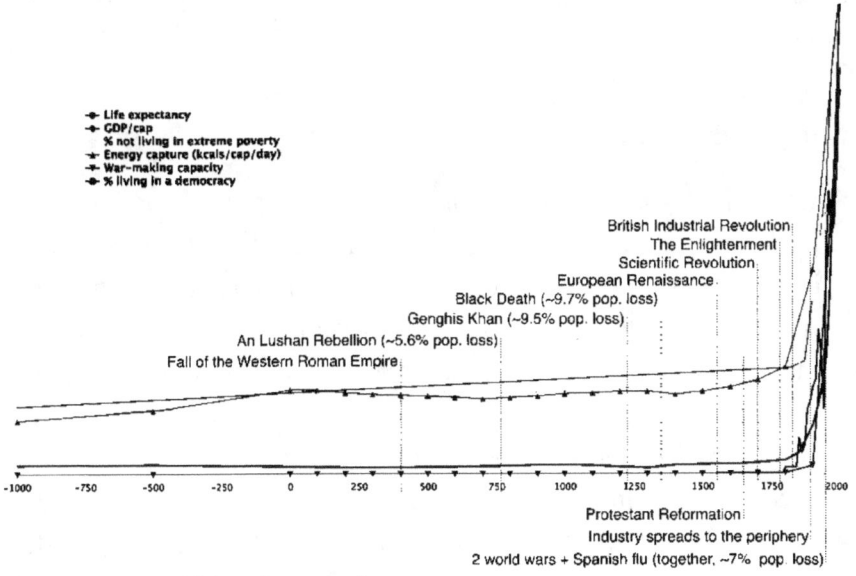

Figure 30 How Life has Improved

Figure 30 demonstrates how man's progress in virtually all measures was essentially flat from prehistory until the Industrial revolution, with only slight improvements in poverty and life expectancy. Yet, when the revolution comes, the changes are dramatic and immediate.

Not surprisingly, these improved living conditions spread worldwide. The world population grew to numbers that in the 1970s and 1980s, the pundits said would be fatal and unsustainable. The predictions were that a growing population would eventually crash the system leading to starvation and collapse. Yet, the production capacity continues to grow even while the population grew unheeded.

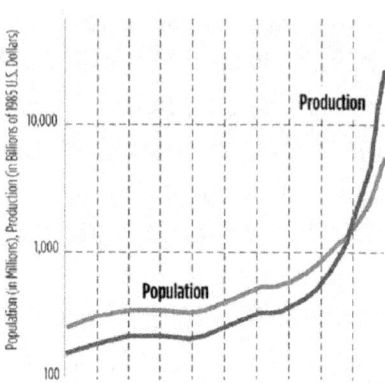

Figure 31 World Population and Production

Life Expectancy

Starting as far back as prehistory and going up until the industrial revolution, global life expectancy hovered around 30 to 35 years of age. The measure of life expectancy can be a helpful measurement of the success of a community. However, it can be a complex measure to work with as it measures the balance between infant and early deaths against those who live until their senior years. Considered differently, the measure of life expectancy can be viewed as the combination of child and infant mortality with the extent of life span. There have always been those that live into their 60s and 70s, although the 90s and 100s are a more recent phenomenon. In 1950, approximately 1 in 67,000 people in America reached their 100th birthday. Today that number is 1 in 6000 people. Although there are a lot of factors affecting how long someone may live, for anyone living today, they have even odds for living to be at least 90, which is impressive when considering that as recently as 1940, the Census Bureau thought those above 100 to be so rare that they weren't worth counting separately.

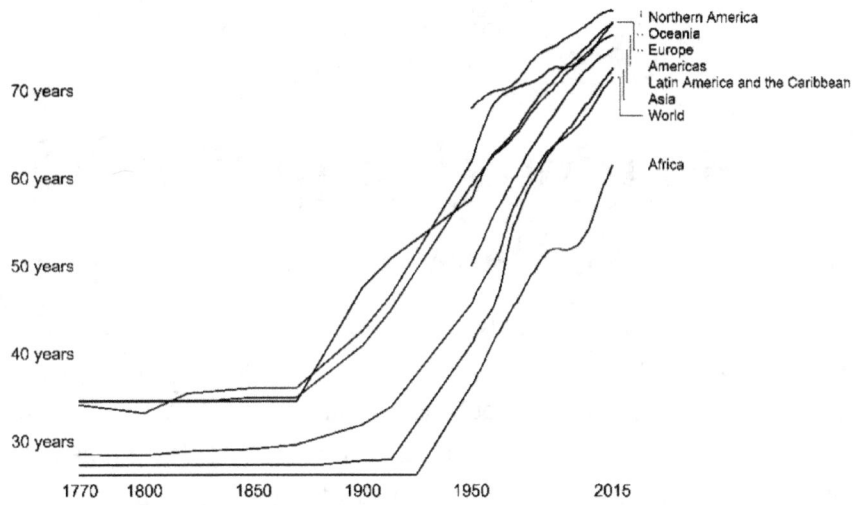

Figure 32 Life Expectancy 1770 - 2015

Primarily, the measure of life expectancy grows when the deaths during early childhood decrease. The child mortality rates within the developed world have plummeted during the industrial revolution due to improvements in food, medicine, and housing. At the start of this dataset, in 1800, children's mortality rate before 5 years was 43%. At the time in Europe, on average, parents lost between 3 and 4 children before the age of 5. The industrial age changed everything. The globalization of the developed world's improvements reduced the global rate to 22.5% by 1950. The further globalization of modern medical practices, strongly driven by petrochemicals, lowered the mortality rate to 4.5% in 2015.[164]

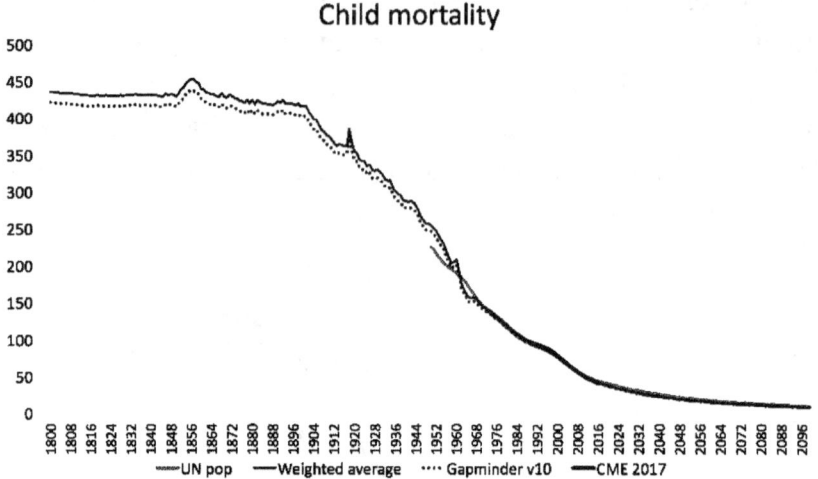

Figure 33 Child Mortality through history (in deaths per thousand live births)

The measure of life span, represented by the age of death for a person who lives beyond the age of five, is the second part of the life expectancy calculation. Population data shows that advances in medicine, safety, and disease control have not only reduced the incidence of infant mortality but mortality at all ages, pushing the peak age of death to ever higher numbers. Advances in vaccines and medical treatment have been so effective that infectious diseases are no longer even in the top-10 reasons for childhood deaths in the developed world. Deaths from congenital anomalies and low birth weight are down over 90% since the 1950s. Overall, about 13 children per 100,000 (0.01%) die between age 5 and 18.[165]

Deaths among those under 50, the children and young adults have dropped dramatically, and the curves have flattened to near zero (see figure 33). Historically, these young adults faced threats to their lives primarily from infectious diseases, safety issues, and war. Technology improvements in the medical field have handled contagious diseases, and efforts in engineering and machinery designs have improved safety. Since World War Two, the dominance of the American military has saved millions from the fate of war. All of which have improved the lives of young adults. Unfortunately, technology has limited influence on human behavior, thus suicide has risen to one of the leading causes of young adult deaths.

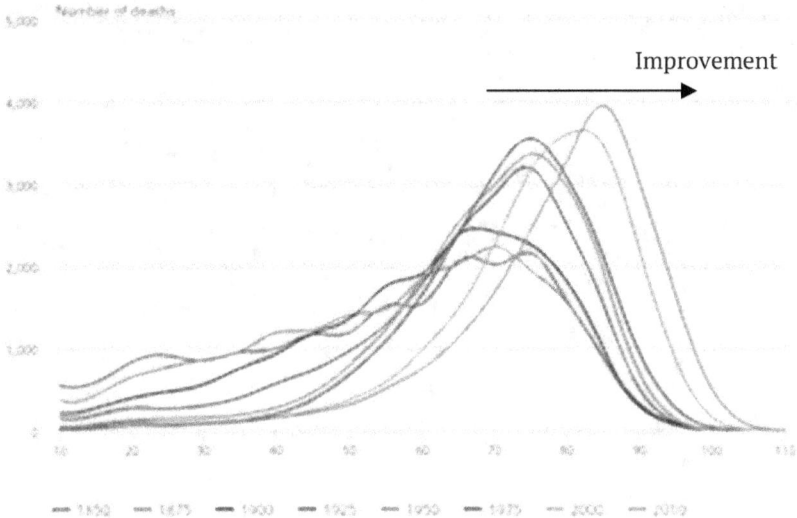

Figure 34 Number of Deaths by Age for UK Males

The improvements in health have continued for middle-aged adults as advances in treatments for heart disease and cancer have reduced deaths. In fact, more and more people are dying within a tighter and tighter range. A study by Alexander Scheuerlien at the Max Planck Institute for Demographic Research showed that the "death spike" (see figure 35) has been getting narrower. In 1933, 50% of American women's age of death was spread across 26 years, while today, that same spike in deaths covers only 16 years.[166] The shift is primarily due to the reduction in deaths occurring in their 40s and 50s, which was once common and now considered an untimely premature death. Today, death before 75 is considered "dying before their time".

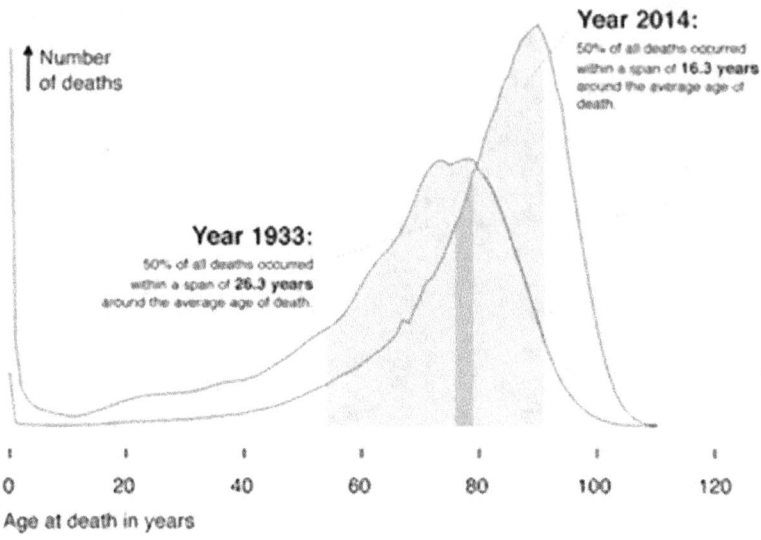

Figure 35 The Death Spike

Much is often made in the media about the fact that the number of deaths increases year to year. But this is to be expected, that while the baby boomer population ages, more of the population will approach the death spike, resulting in an increasing death statistic every year. The industrialization of life and the benefits of fossil fuels allow more of the population to experience life until their most senior years. By the measures of life span and expectancy, life is better with fossil fuels.

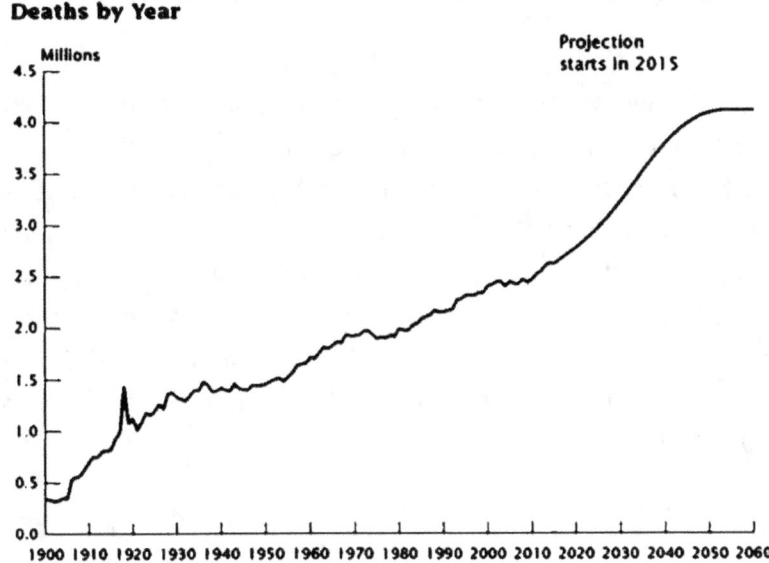

Figure 36 U.S. Deaths by Year[167]

Wealth Generation

As we have seen, before developing fossil fuels, the energy sources that man had at his disposal were limited to human and animal muscle, along with the occasional use of the blowing wind and flowing water. To all but the wealthy, there was little in the way of opportunity or ability to generate independent wealth. The land, which was the means of survival, belonged to the crown or large landowners, leaving the peasants stuck at their mercy with limited opportunities to choose different paths for their lives.

For most of history, money, as experienced in modern society, did not exist. Coinage was generally limited to tokens representing substantial wealth. Today, the cent is considered so unsubstantial that they are found lying around parking lots everywhere. "It's only a penny," so why worry. But when the penny represented four to six hours of labor, the value was considerably different. How does one go from a cent representing significant labor to one where we do not even bother to pick it up from the ground?

What happened was the industrial revolution and fossil fuels, which brought opportunities for wealth to the broader population. With increased opportunities to earn an income came savings, and with savings came investments, all of which creates wealth. In this new modern world, there were now opportunities to become part of the merchant or trade classes and to carve out new opportunities for oneself. Between 1800 and 2010, when adjusted for inflation and other effects, the GDP per capita in the United States grew from $1300 to $31,000.

Figure 37 GDP per Capita, 1500 to 2010[168]

Candidates for political office frequently like to campaign on the idea of a disappearing middle class in America. While they are correct that there is a disappearing middle class that has shrunk from 54% to 41% of the population, it is not for the reason they propose and for which they offer solutions. The candidates would lead you to believe that the missing 13% of the population moves from the middle class into the lower-income classes. However, the lower-income classes are also disappearing, shrinking from 37% to 29% of the population. So, where have the missing 20% of the people gone? They have moved upwards into the upper-class income category, increasing from just 9.0% of the population to 29%. There is roughly the same percentage in the upper-income bracket as in the lower-income one.

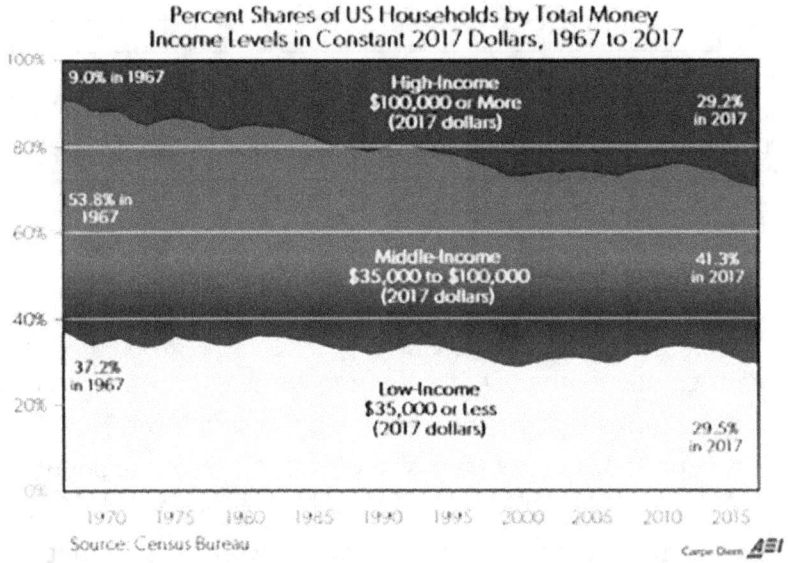

Figure 38 Total Money by Income

A Rising Tide Lifts all Boats. – Sean Lemass

The rising tide of the growth of GDP has indeed risen all boats. When it comes to reviewing extreme poverty globally, it is clear that nothing in history has affected it, like the industrial revolution and fossil fuels. Throughout history, poverty, even what is considered today to be extreme poverty, existed for close to 90% of the world's population, with only the landholders being outside of poverty.

There was only a limited merchant class that would exist between the impoverished and the nobility.

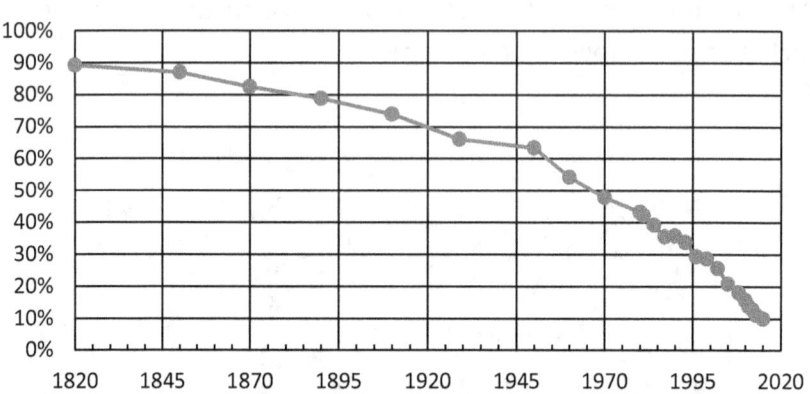

Figure 39 Share of the World Population living in Absolute Poverty

When it comes to quantifying the impoverished, it is necessary to define the terms that delineate what poverty is. America uses the census to periodically set its poverty level at 17% of the population, which is why regardless of the political programs or "war on poverty," the number considered poor always comes out to be roughly 17% of the population. Currently, that level is set at $16 per day per person, leaving 15% of the U.S. population under the poverty line as it was defined in 2010. However, most of those that America calls her impoverished are not so in the global sense of the term. By U.S. standards, two-thirds of the world's population, 5.2 billion people, live in poverty. The World Bank's definition of an "International Poverty Line" is set at less than $1.90 per day.

Historically the worldwide poverty rate ran at 90% of the population. With the advent of capitalism, powered by fossil fuels, the level of unrelenting poverty finally improves to around 80% by 1800. The spread of the technology made possible by primary and secondary energy sources led to immense societal changes, which dropped the poverty rate all the way to 36% by 1990. Over the past 30 years, with widespread access to energy, especially in the form of electricity in India and China, significant gains in the poverty rate have been made, dropping the extreme poverty rate to 7.9%, falling at a rate of 1% per year. There will be an uptick in this rate in 2021 due to lockdowns, while the wealthier middle and upper-income communities could work from home, those whose work required direct interaction with people or relied upon the working population for their incomes faced a significant loss of wages. While the upper-income bracket saw 10% unemployment during the peak of the lockdown crisis, the lowest bracket experienced 40% unemployment. Most of the newly impoverished are in what is considered middle-income countries.[169] By the World Bank standards, the extreme poverty in America is too small to measure, limited primarily to the homeless.[170]

As the developing nations gain access to energy resources, the wealth goes up, and poverty goes down. With energy comes the hopeful end to poverty and a brighter future.

Food Security

The United Nations continues to sound the hunger alarm bells. Their Sustainable Development Goal #2, "End hunger, achieve food security and improved nutrition," aims to eliminate all forms of malnutrition by 2030. In their latest environmental report, "Making Peace with Nature," the IPCC notes, "Over the last 50 years, the global economy has grown nearly fivefold, due largely to a tripling in the extraction of natural resources and energy that has fueled growth in production and consumption. The world population has increased by a factor of two, to 7.8 billion people, and though on average prosperity has also doubled, about 1.3 billion people remain poor, and some 700 million are hungry." While the world's progress has been impressive, they consider that not enough has been done. But are they considering who remains hungry? And what it takes to reach them?

I grew up in an era in Middle America where we heard the news of the starving children in India. As Dinesh D'Souza is fond of pointing out, Americans today are no longer worrying about those starving children. Instead, America is busy worrying about the millions of newly educated, Indian middle-class youths who are moving into the workforce. In the past 50 years, India has moved from being on the verge of starvation to being a food exporter, producing more food than it needs.

Dr. Borlaug's green revolution has fed the world in unprecedented ways through better crops and better fertilization. According to a Department of Energy report, the process of using natural gas to create fertilizer is "referred to by some as the most important technological advance of the 20th century.... Between 3 and 5 percent of the world's annual natural gas production – roughly 1 to 2 percent of the world's annual energy supply – is converted using the process to produce more than 500 million tons of nitrogen fertilizer, which is believed to sustain about 40 percent of the world's 7 billion people. Approximately half of the protein in today's humans originated with nitrogen fixed through the Haber-Bosch process."

Globally, the number of malnourished people has dropped from 37 percent in 1970 to just 8.9 percent in 2019.[171] An impressive reduction which was accomplished while reducing the amount of land dedicated to food production. Currently, the world produces enough food to feed 1.5 times the global population, or roughly 10 billion people.[172] The primary reason for food wastage is the lack of refrigeration in developing nations, which could be solved with reliable electricity.

Ironically, organizations like the United Nations Food and Agriculture Organization (FAO) predict severe food shortages by 2050 as the population grows to 9.1 billion people before leveling off. A 2019 study indicated that farmers would need to produce 70% more food, a total of 1 billion additional tons of wheat, rice, and other cereals.[173] Why the need for extra food when there is already enough to feed 10 billion? Their unstated premise is that those 9.1 billion people will be leading better lives. Lives in which they will demand modern food services and choices. Lives that will consume more calories daily than they do today. The researchers expect that a greater portion of the 9.1 billion population will live in the world developed by the use of fossil fuels. It is projected that if the improvements in wealth continue, especially in Asia, the percentage of the world population living in the developed countries could reach 50%, irrespective of mass migrations. All of whom would expect the appropriate developed world lifestyle.

Saving Lives

As societies become more affluent, they become safer. As Ronald Bailey notes, "Richer societies are likely reducing their weather losses by establishing better early warning systems, enacting stronger building codes, and constructing firmer levees. People may be protecting themselves ever better against the consequence of storms and floods, even though the weather is getting worse."

Over the past century, the number of worldwide deaths from climate-related events has fallen by 98%.[174] Deaths due to natural disasters, including drought, floods, storms, and earthquakes, peaked around 1920 and have been falling ever since. This is not to say that the cost of damage has been reduced, as the IPCC's 2014 Adaption report indicates, "Economic costs of extreme weather events have increased over the period 1960-2000, with insured losses increasing more rapidly than overall losses."[175] With society's growing wealth, property values increase, and more "things" and more expensive "things" get damaged.

The reduction in deaths can be attributed to technology as improvements in weather forecasting, information dissemination, and transportation enable people to get out of the way before significant storms hit. It is estimated that 8000 deaths resulted from the 1900 Galveston hurricane, and 235 people, primarily children, died from the Schoolhouse Blizzard of 1888, essentially due to the lack of forecasted warnings. Since the development of technology, it is known days in advance of any hurricane or blizzard possibility. Technology now provides forecasted temperatures within a degree ten to fourteen days in the future. That kind of forecast could have saved many lives in the past. As an example, as the developing nation of Bangladesh has implemented new technology into their forecast and warning system, a country that once lost 500,000 people in a single flooding event in 1970, today reports people displaced by the annual monsoon seasons, not the number dead. The nations' new technology was funded largely by the country's ever-expanding clothing industry, which in turn is enabled by the creation of synthetic fibers created from petrochemicals.

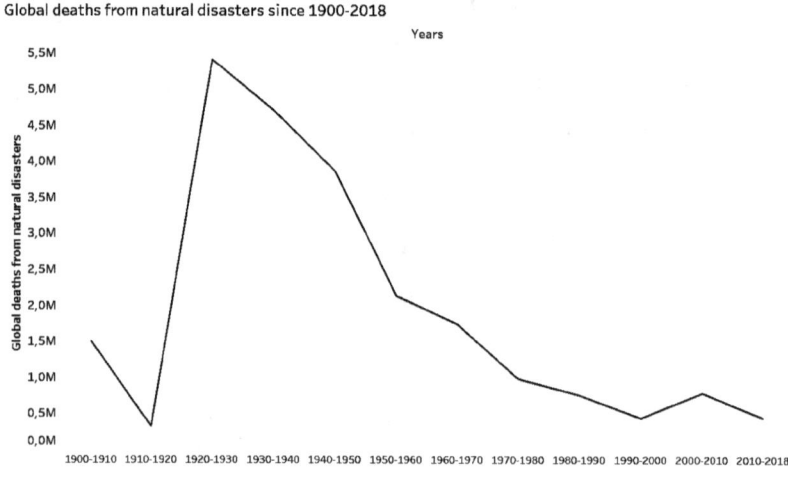

Figure 40 Deaths From Disasters

A National Highway Transportation and Safety Administration (NHTSA) study finds that since 1960, more than 350,000 lives have been saved by introducing seatbelts and another 52,000 from the introduction of airbags, not counting all those whose injuries were greatly reduced. The savings from the introduction of collision avoidance and autobraking systems will never be countable as they would have been the result of accidents that never occurred. The wealthier the country (High SDI), the lower the rate of deaths from road accidents, largely due to introducing the newest technology. While declining for everyone, equipping the least prosperous nations with more of the latest technology would reduce the rate everywhere. The poorest countries suffer from the poorest roads, although China's Belt and Road programs have built better roads, which have resulted in higher travel speeds, which will result in more deaths unless the vehicle technology improves.

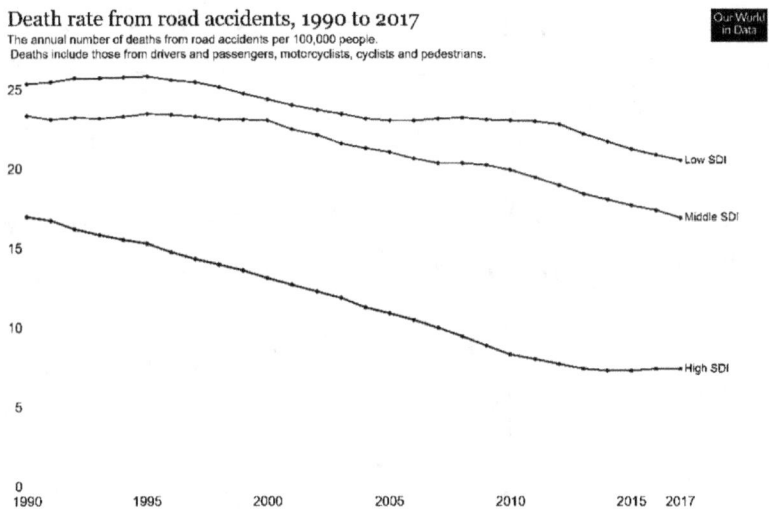

Figure 41 Death Rate from Road Accidents by Socio-demographic index

Every incident which does not happen because technology intervenes, whether automobile, aircraft, ship, subway, trains, or even pedestrian, will never generate a headline or news story, so it is easy to discount the effect technology had. Every day, however, lives are saved and made safer by technology powered by reliable energy.

Liberated Lives

Robert Bryce describes in his book, **A QUESTION OF POWER: ELECTRICITY AND THE WEALTH OF NATIONS**, that "without electricity, women must pump water, heat water over a fire or a stove, use the heated water for washing or cooking, and constantly tend the fire or stove." With reliable energy, everyone can be liberated from so much drudgery. As Meredith Angwin comments in **SHORTING THE GRID**, "When you have enough electricity to run a washing machine, women become empowered. When you can run a washing machine, you have "real electricity." And you are probably connected to the grid." From her book, she provides a plot of births per woman as a function of GDP per capita, as seen below. The fact is that with more energy, people live better lives. When their lives are better, women become more independent and educated, resulting in lower birth rates.

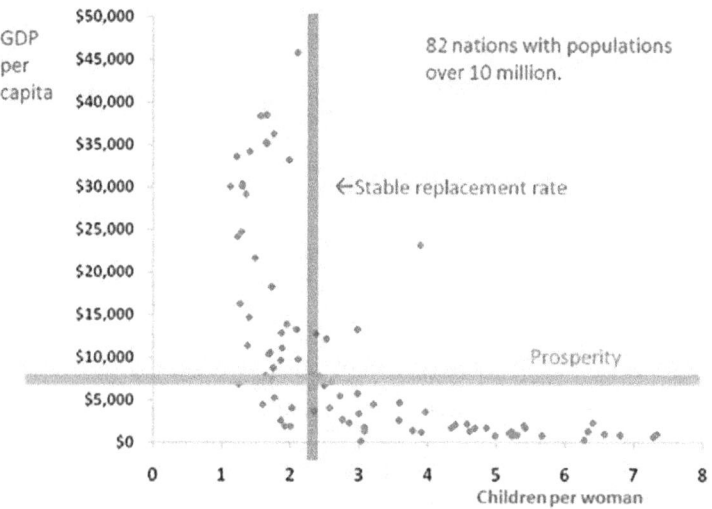

Figure 42 Births per Woman by GDP

The proportion of global energy derived from biomass fuels has fallen from 50% in 1900 to around 10% today. However, nearly 90% of the rural populations in the developing nations still rely on unprocessed biomass in the forms of wood, dung, and crop residues. Exposure to indoor air pollution caused by the burning of these substances results in increases in lung cancer among non-smokers and persistent asthma and respiratory distress among the women and young children who spend most of their time indoors.

China faced a complicated problem in the 70s and 80s when residents of rural villages had to choose whether to use the animal dung as fertilizer on the fields or burn it for warmth and cooking. It was easy to choose the immediate need to ward off the cold of winter over the growth of next year's crops. The result was crop failures due to a lack of fertilizers. The lack of fossil fuels has put these farmers at risk for their health.

As John Tamny wrote in **WHEN POLITICIANS PANICKED**, "this wasn't always the case. Before automation, and yes, global trade, the vast majority of human exertion was directed toward the creation of food. Work was life, albeit not in a happy way. People worked dawn to dusk six days per week in order to survive. Most worked until they died."[176] Whether liberated from the drudgery of work within the home or in the farm fields, the time saved is put to further use by the reduction of time spent in journeys when motorized transportation replaces animals. All that time saved is used for leisure, enjoyed with the benefit of petrochemicals.

It is easy today to take for granted the benefits which fossil fuels bring to modern society. Life is longer, wealthier, freer, and safer due to the technologies which fuels and electricity have enabled. The **Green Solution's** desire to eliminate fossil fuels put at risk all the gains that humanity has enjoyed. The advances in child mortality, extreme poverty, hunger, and personal safety all hang in the balance, dependent on the actions of those who make policy. As

Robert G. Brown wrote in a letter to the Editor for the Wall Street Journal, "Universal poverty isn't an acceptable solution to any problem that I can think of."

8

The True Story of Renewables

> "Without a doubt, the biggest problem with mainstream renewable energy is intermittency. Wind power is only generated when it's windy; solar power is only generated when sunny. This creates several fundamental issues."[177]
>
> *The Green Age UK*, August 2019

Valentine's day 2021, and winter storm Uri swept through the country, affecting Texas and 36 other states. The storm, lasting from February 13th to 18th, revealed the fundamental weaknesses in Texas's electrical generation system. The wintry weather and near-record lows created record demand for electricity, the failure of which left 4 million Texans without power and 57 dead.[178] Total monetary losses from the power failures associated with the storm exceeded $20 billion, including 14 million gallons of milk that could not be pasteurized, 1.6 million chickens that could not be kept warm, and the loss of millions of dollars of tomatoes, lettuce, and other crops. Once proud of their electrical grid's independence, Texans demanded answers to how a state built around the energy industry could fail to have sufficient power for a winter storm.

The Electricity Reliability Council of Texas (ERCOT), responsible for maintenance and management of the Texas power grid, reports that as of that date, Texas electrical generation capacity was primarily from natural gas (51%), coal (13%), and nuclear (5%), along with renewables of wind (25%), solar (4%) and Hydro and biomass making up the remaining 1.9%. In recent years, the Texas grid has been retiring coal facilities while installing wind power. The wind generation capacity has grown from 11% of the grid in 2015 to the current 24.8%.

Many problems contributed to the Texas grid's failure during winter storm Uri, and most were not directly related to the renewables on the grid. However, that is not to say that the issues were not indirectly related to the push to install renewables, doubling the number of wind turbines in the past five years. While at the same time, Texas has shutdown 11 coal-powered plants, with a total retired capacity of 6.2 Gigawatts. There remains 17.4 GW of coal capacity, although half of that is already scheduled to close.[179]

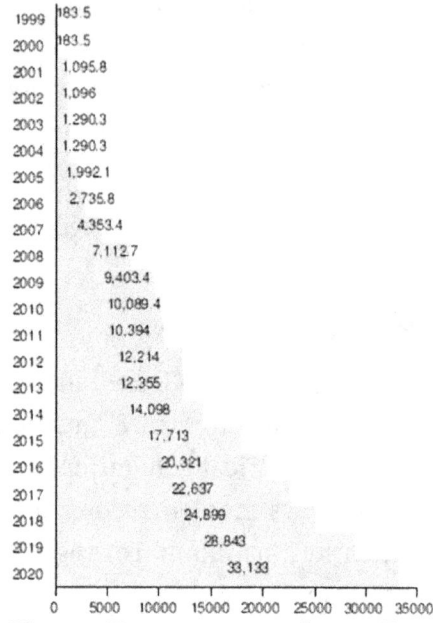

Figure 43 Megawatts of Installed Wind Generation Capacity in Texas

The primary source of the Texas grid's power outages during the winter storm stemmed from demand continually outstripping supply, leading to the grid having difficulties maintaining voltage and frequency standards.

Circuits within your home are protected by circuit breakers or fuses, so damage would be limited if something terrible were to happen. Likewise, when the power generation facilities recognize that the grid's voltage and frequency are not within prescribed limits, the facility trips to protect the millions of dollars of equipment tied to the grid. This is ultimately what happened as some 185 generating units tripped offline during the storm.[180] ERCOT's review of power generation during the event indicates that the coal and nuclear plants successfully provided the baseload, their constant generated output. At the same time, the renewable contribution shrank considerably, which left the natural gas plants to make up the difference in increased demand. The gas-powered plants needed additional fuel to provide more electricity, while natural gas heated homes needed extra gas to keep warm. Thus, there was a lack of gas to go around to everyone, limiting the gas-powered generators' output. While the power generated met the forecasted needs as anticipated by ERCOT, it was insufficient to the actual demand.

Much was made in the news about the problems with the wind turbines during the storm. While these issues might be new to the general public, they were well understood ahead of time. Several factors influenced the ability of the wind turbines to operate in extreme cold.

1) The designed temperature range for the most common wind turbines based on the manufacturer's specification is -20° to +45°C (-4° to 113°F). Temperatures in the western Texas panhandle reached -22°F overnight on February 16, 2021,[181] exceeding the minimum operating temperature specifications, requiring their shutdown to protect them.
2) Wings need deicing. Any airline passenger who has traveled during winter is aware that when the conditions warrant, every airliner is sprayed with deicer fluid before takeoff due to the loss of lift associated with ice on the wings. Currently, Wikipedia lists 32 major airline crashes due to icing of the wings.[182] Deicing of wind turbine blades is accomplished by helicopter in a slow, expensive, and laborious task. Which is

not an activity that can be done in the middle of a winter storm.

The grid authorities are fully aware of these issues and have designed their generation plans around the anticipated availability of wind and solar systems. In their winter fuel type planning, ERCOT estimated that of the wind generation capacity, 43% would be available along the Texas coast, 32% in the Texas Panhandle, and 19% for all other state regions.[183] Likewise, ERCOT reports that during the winter, solar power, representing 4% of total grid capacity primarily through roof-mounted panels both residential and commercial, can be expected to only provide 7% of that 4% to the electrical grid for usage[184] due to low angle sun, shorter periods of daylight, and coverage by snow. As ERCOT predicted, as the storm arrived, the availability of solar and wind declined. In actuality, the renewables performed better than expected during the peak of the storm. However, it was still insufficient to make up for coal facilities that had been shut down. The biggest issue with the renewables was the rate at which their generation power was lost. The weather was mild at midday on February 8th as the storm front approached. A point in time where the renewables were generating nearly 80% of the demand (see Figure 44). As the clouds rolled in and the temperatures fell, so did the wind and the sunshine. The renewable power suddenly vanished. The need for gas-powered plants to make up the difference was nearly instantaneous; however, gas-powered plants take time to come online, leaving a gap in the energy supply-demand equation. And with the gap came disaster.

As Meredith Angwin points out in her book, **SHORTING THE GRID**, "Nevertheless, despite the reality of laws of nature—the sun doesn't always shine, the wind doesn't always blow, and inverters don't [always match] the grid—legislators make ... laws saying their state grid must be 100% renewable. The laws of nature are not repealed by these renewable-mandate laws, and yet the laws are passed."[185]

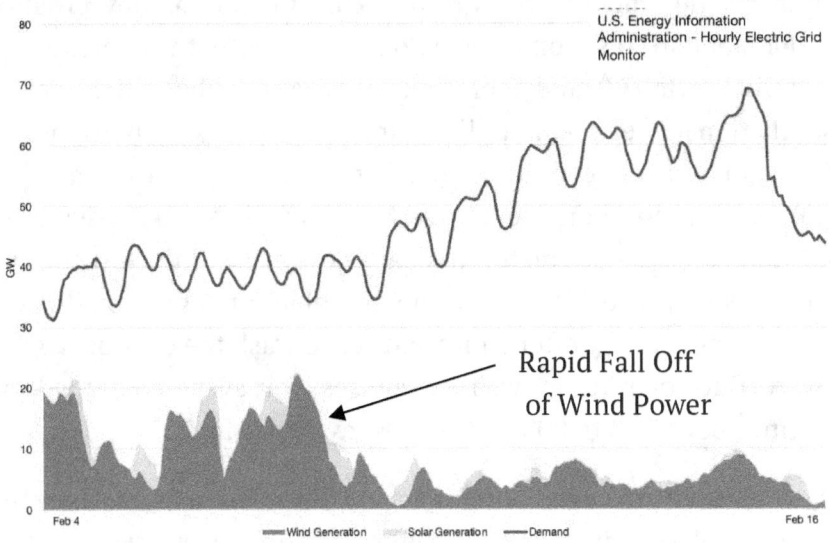

Figure 44 ERCOT Generation and Demand Feb. 4-16, 2021

Solar Shines Dimly

From the 1975 introduction of CAFE (Corporate Average Fuel Economy) standards until the improvements in the testing methods in 2008, it was a running joke about just how unrealistic the advertised automotive efficiency numbers were compared to real-world driving. It was not unusual for the numbers to be off by 20-30% from what drivers would really get. The reason for the discrepancy was that the test the EPA was using was a set profile of speeds, starts, and stops while on a dynamometer test stand. Due to this setup, there were no hills, no headwinds, unrealistic start/stop/idle times, and the test simply did not reflect the driving profile that most drivers encounter. In other words, it was an overly optimistic measure of fuel efficiency.

Likewise, the rating on a solar panel is an overly optimistic measure of performance. As stated on EnergySage.com, "All solar panels are rated by the amount of DC (direct current) power they produce under standard test conditions. Solar panel output is expressed in units of watts (W) and represents the panel's theoretical power production under ideal sunlight and temperature conditions."[186] The standard test conditions (STC) for a solar panel are at 77°F (25°C), an atmospheric density of 1.5[4],[187] and with an irradiance of 1 kilowatt per square meter of solar energy (max full sun) shining on the panel.[188] At high noon, this illumination level would occur with a cloudless sky after a rain shower to wash the dust out of the air. As one solar design website advises, "on average, you will be getting about 75% to 80% of the power you pay for."[189]

Unfortunately, the news doesn't get any better from there. Once the solar panel gets installed, it begins to suffer from two specific problems. Firstly, the panels need to be cleaned regularly. A report by the U.S. National Renewable Energy Laboratory reports that "moderate soiling that results when panels are not cleaned monthly is generally calculated to result in a 30% energy yield loss per year, while longer-term, cumulative soiling cementation of dust materials can result in a 100% loss of yield."[190] The standard installation of solar panels on rooftops provides a challenge to such monthly cleaning, often with leased panel contracts requiring hiring specific cleaning crews. Secondly, solar panels are not unlike most equipment; they don't run at 100% and then die. Instead, they degrade by 1 to 3% per year due to the effects of weather, microcracking, and UV light. Solar panels will continue to generate power for a long time, but such installations are considered to have a lifetime between 10 and 20 years due to weathering and

---

[4] A measure of the optical path through the atmosphere. Atmospheric Density = 1 is on the equator at high noon at sea level. Density = 1.5 is equal to 500' about sea level.

deterioration of their physical construction resulting in a reduction of power output.

As part of a sales pitch, the Energy Sage website provides information about installing solar panels, including calculating the output from a panel installation.

> *For the sake of example, if you are getting 5 hours of direct sunlight per day in a sunny state like California, you can calculate your solar panel output this way: 5 hours x 290 watts (an example, wattage of a premium solar panel) = 1,450 watts-hours, or roughly 1.5 kilowatt-hours (kwh). Thus, the output for each solar panel in your array would produce around 500-550 kWh of energy per year.*[191]

It is vital to note here that in a sunny state like California, they expect that you would receive about 5 hours of usable sunlight. The amount of sunlight received will vary significantly by location. In table 7, each state is listed with the number of clear days and the average percentage of sunshine. If you can only typically expect 5 hours in California with their 68% sunshine, then clearly solar panels will be less effective in Vermont, Oregon, or Washington with less than 50% sunshine and so few clear days.

Table 7 Average Annual State Sunshine (DE and WV data missing)[192]

State	%Sun	Clear Days	State	%Sun	Clear Days	State	%Sun	Clear Days
AL	58	99	LA	57	101	ND	59	93
AK	41	61	ME	57	101	OH	50	72
AZ	85	193	MD	57	105	OK	68	139
AR	61	123	MA	58	98	OR	48	68
CA	68	146	MI	51	71	PA	58	87
CO	71	136	MN	58	95	RI	58	98
CT	56	82	MS	61	111	SC	64	115
FL	66	101	MO	60	115	SD	63	104
GA	66	112	MT	59	82	TN	56	102

HI	71	90	NE	61	117	TX	61	135
ID	64	120	NV	79	158	UT	66	125
IL	56	95	NH	54	90	VT	49	58
ID	55	88	NJ	56	94	VA	63	100
IA	59	105	NM	76	167	WA	47	58
KS	65	128	NY	46	63	WI	54	89
KY	56	93	NC	60	109	WY	68	114

Furthermore, those sunshine hours are not all created equal. Overhead sun has less atmosphere to shine through and therefore is brighter and more energetic. As the sun rises into the sky and then sets, the efficiency changes constantly. Unless the panels are equipped with motor drives to follow the sun, which adds considerable complexity and cost to any installation but has the benefit of increased hours of higher efficiency. Unfortunately, the power used by those motor drives reduces the gains from the increased panel efficiency, offsetting a high percentage of the gains.

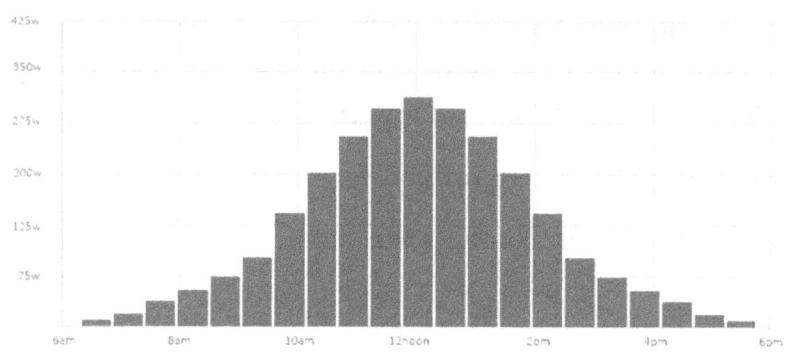

Figure 45 Solar Panel Output by Time of Day[193]

The standard test conditions for solar panels were carefully stated as "with a cloudless sky after a rain shower to wash the dust out of the air," as particles and pollution in the air will diminish the amount of light energy that gets to the surface of the panel. A 2017 study of system performance across India, China, and the Arabian Peninsula showed that air pollution could result in an additional 17 to 25% reduction in energy generation. Overall, haze, smoke, dust, factory soot, or even intense pollen clouds all hurt the panels' ability to operate.

Solar may operate well in the southwestern deserts and other places where the sun shines in pollution-free environments, while in other areas such as the smog of Beijing or the pacific northwest's cloudy weather, it will fail to be as successful. Complex calculations would be required to factor in all the lifetime variations, along with the location, pollution, and time of day variables. Just as cars never got the mileage advertised except under particular circumstances, even by taking extreme care and with a bit of luck, solar panels will never achieve the rated power.

Wind Blows Lightly

The United States has 122 gigawatts of wind power capacity installed but generates significantly less power than that. The size of the capacity is only part of the equation of how much power a wind farm can generate. In theory, if all the wind power systems worked at total capacity for a full 24 hours, it would be possible to generate 3000 Gigawatts-hours (3 million MWh). The record generation as of the end of 2020 was set on Dec 23, 2020, at 1.76 million MWh (roughly 60% of capacity), which accounted for 17% of that day's electrical generation.[194]

In 1919, the theoretical limit of the power that could be generated by a wind farm was calculated as a maximum of 59% of the energy of the wind. This value, the Betz limit, represents the maximum possible extractable energy for any wind farm design.[195] In reality, as that is a theoretical limit, wind turbines capture around 40% of the potential wind energy, although some designs are more efficient than others. In addition to the blade efficiency, there is mechanical efficiency that gets involved in the power equations. The most successful wind turbine in the market is the Siemens 3.4 MW horizontally mounted three-bladed unit. This unit is rated as capable of generating 3.465 Megawatts of power at wind speeds between 15 and 25 meters per second (33 to 56 mph). At lower wind speeds, which could be a significant portion of the time, the power generation is proportionally lower, reaching zero output at around 8 to 10 mph wind speed. The requirements to find those sustained 33 to 56 mph winds mean that the units' placement is severely restricted, limiting the number of units that can ultimately be placed.

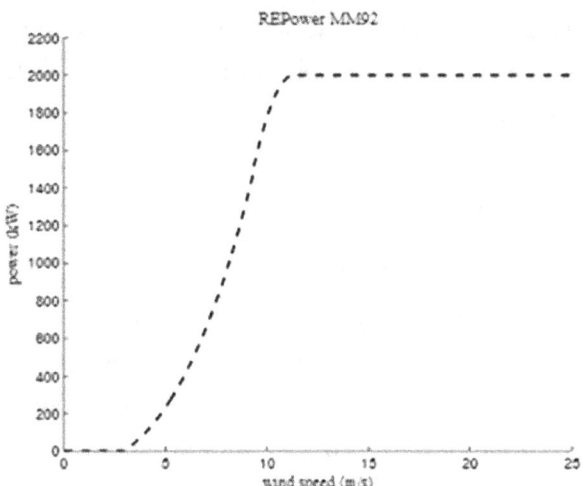

Figure 46 Wind Curve for Siemens Wind Turbine

The second variable involved in wind turbine generation is the capacity factor, which is the average power a unit expects to output divided by the maximum rated power capacity. Depending on the turbine's placement, typical weather conditions, and daily and seasonal wind variations, this number varies between 26 and 52%. The average for all the units installed before 2020 was 35%.

Unit efficiencies have been improving over time, so the following chart is lower than the current number as the data covers 2001 to 2013, but the trend is correct. Nationally, nature provides wind that peaks in April with a smaller peak in November; however, it is lowest during July and August.

The Energy Information Administration breaks down this information by region, which shows a similar pattern for all the regions except for the west coast, which does not experience the summertime dip. Here in Texas, however, we experience a significant drop in the summertime, just when the peak in cooling is required.

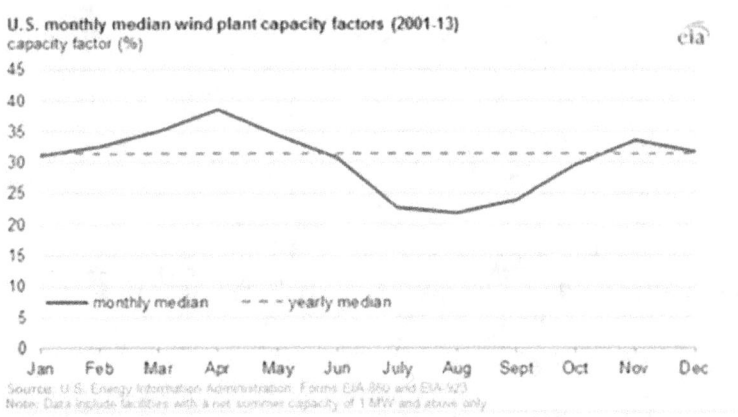

Figure 47 Anticipated Wind Turbine Plant Capacity

In addition to the seasonal variation of wind, there are, of course, hourly variations. Figure 48 (following) shows the wind generation over a single week of May 2020. Wind's ability to meet the demand fluctuates across the day, and worse yet, it is offset from the peak demands with the wind peaking at night.

In May 2020, The Jersey-Atlantic Wind Farm put out an updated fact sheet on their wind project, which indicated that in 2005, they installed 5 wind turbines, each rated at 1.5 Megawatts for a combined total of 7.5 Megawatts. The project aims to provide 2.5

Megawatts of power to the Atlantic County Utilities Authority to process wastewater. Having failed to take the 30% capacity factor into account, they have been forced to change the documentation from "all of the electricity required to run the plant can be generated by two of the five wind turbines" to "provides up to 60% of the plant's electrical needs."[196] A significant change, they had expected 2 units to provide 100% of the needs, and now all 5 units are meeting only 60% of the requirements. Care must be taken in determining just how much real power wind will generate and when.

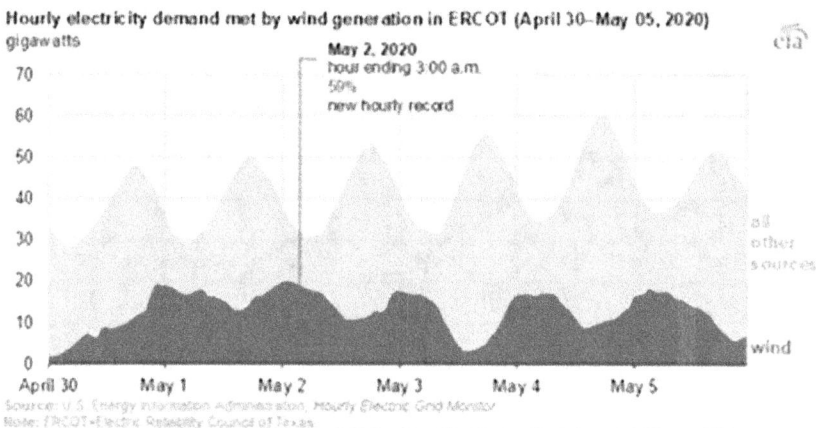

Figure 48 Wind Generation Creation

Birds are killed and injured every year due to collisions with powerlines and facilities. Unfortunately, the American Bird Conservatory studies show that an estimated 1.17 million birds are killed annually by wind turbines in the United States.[197] Due to the wind farms' typical remoteness, the additional power lines required to connect the turbines will increase the number of powerline bird deaths. As much as the aviary toll that the windmills cause is horrifying, the human risk is real. Every year 3800 blades fail and become flying projectiles, covering distances as much as 0.8 miles from the turbine mount. In 2019, Iowa got 42% of its power from the 5000 wind turbines in their state. But in October 2020, the state was forced to take nearly half of its turbines offline when failed blades revealed possible safety issues. The turbines were forced to remain out of service while "full analysis and inspections" could be completed.[198] The website "stopthesethings.com" catalogs many turbine collapses, blade failures, and nacelle fires for those interested in reviewing these things. While collecting worker injury data is difficult, news reports exist for at least 70 workers who have died from falls or injuries sustained while atop the towers. To say that wind turbines are safe is being simplistic, as the nearly 4000 blade failures per year represent nearly 2% of all installed blades failing in a given year. This may well explain why you see so many wind turbine blades traveling on the roads.

Renewables that are not all that renewable

Liebig's Law: growth in any activity is dictated not by total resources available but by the scarcest resource (limiting factor).

The question of whether a power generation system is genuinely renewable energy is whether the system can renew itself. Once the world has transitioned to 100% renewable energy, there will still be the demand for both additional units to expand the available power and replacements for installations that break or wear out (more on that later). While renewables can generate electricity to operate motors, turn on the lights and run all our electronic gadgets, they cannot provide the materials which were required to build the units in the first place.

According to their website, the commonly installed Siemens Gamesa wind turbines use three custom-built blades constructed from Fiberglass reinforced polyester with epoxy resin.[199] The main structural element is an inner girder built from "fiberglass and carbon pre-coated with epoxy resin — a thermostable polymer that hardens when mixed with a catalyst agent."[200] Two half-shells constructed of fiberglass are bonded together with yet more of the epoxy polymers, cured and then painted. So, the blades themselves are primarily built from materials derived from petroleum as the original feedstock. On average, the replacement of turbine blades every 10 years is common practice due to damage and improved technologies.

Further, the typical 3.4-Megawatt Siemens wind turbine requires 60 gallons of lubricating oil which must be situated alongside the gearbox, as much as 300 feet in the air. Typically, this oil must be changed every 8 to 12 months by crews who climb atop the structures.[201] In particularly dusty or dirty locations, the oil may need to be changed more often. Because turbines may be located in saltwater, dusty, high- or low-temperature environments, several specialized lubricants have been developed for this use. In extreme weather locations, two oil changes are needed each year, lighter oils for winter, heavier oils for summer. But one thing is for sure, the wind turbine can't create the lubricants it needs to be self-sustaining.

Wind turbine construction requires other raw materials that cannot be produced by the wind turbines electricity alone: aggregates and crushed stone (for concrete), bauxite (aluminum), clay and shale (cement), coal, cobalt (magnets), copper (wiring), gypsum (cement), iron ore (steel), limestone, molybdenum (an alloy in steel), rare earths (magnets; batteries), sand and gravel (cement and concrete), and zinc (galvanizing).[202] It is estimated that each Megawatt of wind power requires 103 tons of stainless steel, 402 tons of concrete, 6.8 tons of fiberglass, 3 tons of copper, and 20 tons of cast iron.[203] While at least a reasonable number of installations could be manufactured by fossil-fuel powered industries, each use of those fossil fuel processes cannot be done with wind-generated electrical power.

Since solar panels are not such large structures, it may seem believable that it is more likely to be self-sustaining. Depending on the solar technology being used, it may require all or some of the 17 minerals listed: arsenic, bauxite, boron, cadmium, coal, copper, gallium, indium, iron ore, molybdenum, lead, phosphate, selenium, silica, silver, tellurium, and titanium dioxide. The best source for these materials would be, in fact, old solar panels. It is estimated that 314 kilograms of CO_2 are generated for every square meter of solar panels built, roughly half that per kilowatt-hour of fossil fuel electricity.

As Heartland Institute Analyst Paul Driessen noted that without fossil fuels, it is likely that "we couldn't mine the needed ores, process and smelt them, build and operate foundries, factories, refineries or cement kilns, or manufacture and assemble turbines and panels. We couldn't even make machinery to put in factories... [Renewables] cannot provide the power required to manufacture turbines, panels, batteries or transmission lines."[204]

Hydroelectric Power

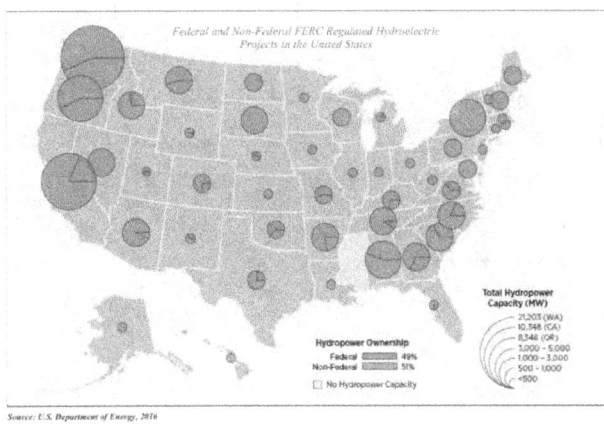

Figure 49 Hydroelectric Power Sources

Using the flow of a stream to provide power goes back into the antiquities of time. The water wheel's use to grind grains and pump billows was typical throughout the middle ages into the early industrial era. That changed from gaining mechanical to electrical energy from the moving water in 1881 with the Schoellkopf Power Station No. 1 at Niagara Falls, NY. Multiple stations for power were built at the site between then and 1924, the ruins of which can still be toured.[205] By the end of the 1940s, hydroelectric projects would provide as much as 40% of the nation's electricity needs.[206] The Civilian Conservation Corps building boom built many of the giant projects from Hoover and Grand Coulee Dams to the Tennessee Valley Projects.

Present in every state except Mississippi, the total installed base of hydroelectric projects provides a combined 101 Gigawatts of power, representing 8% of the electrical generation in the United States and 48% of all the renewable generation. The projects are roughly split 50/50 between federal government ownership (operated by the Corps of Engineers) and state and private ownerships (regulated by the Federal Energy Regulatory Commission).[207]

No significant projects have been undertaken since the 1970s due to the cost of such large projects, running out of untapped water flows, and a general resistance by local residents to constructing the lakes and reservoirs needed for the dams to operate. We can see the importance of utilizing this power from such natural sources to prosperity since the developed world already gains power from 70% of their potential sites. In comparison, the undeveloped world utilizes less than 3% of theirs, so this represents a significant untapped resource.

The Power of Tides

Wind and solar always suffer from the unpredictability of intermittency, which would never be the case for power generated by the tide's constant predictability. The earliest designs for capturing the tides' energy were done by Archimedes (287-212BC) and Da Vinci (1452-1519). Research into tide power has been a hot topic for consideration from at least the 1970s. However, while numerous proposals have long floated around on capturing the tides' energy, limited construction has actually been undertaken. A couple of experimental projects in South Korea and France produce approximately 250 MW of power each.[208]

Any ocean-based power source has an inherent flaw, corrosion. Read any of the numerous books about sailors traveling by sailboat non-stop around the globe, and you will soon realize that the sailors spend a majority of their time fighting the corroding effects of saltwater. Forget engine maintenance for a few days, and the repair task may not be possible. Saltwater has a way of corroding everything which it contacts. Add to that invasion by various marine life and keeping units running efficiently under the ocean will be a challenge. Applicable where the tides are strong, but they do risk damaging the tidal basin's ecology where they are constructed. Time will tell if more of these systems will be built.

Energy Density

It is called the rocket problem. For every pound of payload you want to launch into space, you will need a certain amount of energy in the form of fuel to accelerate that weight off the ground. When you add just another little bit of weight, then you must add more fuel. And when you add more fuel, you need yet more fuel to launch the fuel you just added. Because of this dilemma, rockets will often have 85-95% of their weight just to be fuel. For the Saturn V Rocket, 5.6 million pounds of the total 6.7-million-pound rocket was fuel or about 84% of the total weight. To just clear the tower (345 feet), the Saturn V rocket burned 4% of its fuel, a distance that was 0.00000008% of the trip.

Clearly, it does not take a rocket scientist to know that a rocket scientist needs to use a fuel that gives him the most push for every pound of fuel used, and preferably, that pound of fuel would take up a minimum amount of space in the rocket, after all, less space means more pounds of fuels, means more oomph! The density of the energy would be measured as pounds per cubic foot (metric: kg/m^3)

If the Energy equals oomph per pound and Density equals pound per volume, then Energy Density equals oomph per volume.

Now, what was the purpose of this excursion into rocket design?

Take any automobile, and you have a fuel tank of a fixed size. It might be 12 gallons or 15 or 18, but the volume is fixed. Now, when that tank is filled, fuel's energy density in that fixed volume tells you how much power you will have to burn. Among typical fuels available at the pump, diesel has the highest density at 128 KBTUs / gallon (which is why the super high mileage cars are diesel), gasoline second at 116 KBTUs/gallon, and lastly ethanol at 76 KBTUs / gallon (which is why your gas mileage goes down with 10% ethanol).

Table 8 Energy Density by Fuel

Fuel	Energy Content (BTUs / Gallon)
Low Sulfur Diesel	128,488
Biodiesel (B20)	126,700
Biodiesel (B100)	119,550
Gasoline	116,090
Ethanol (E85)	88,258
Propane	84,250
Ethanol (E100)	76,330

Ethanol isn't the Answer

The largest potential risk to food security in a fossil fuel free age may be the increased growth in the use of crops for biofuels to power long life, lower turnover vehicles. While providing nearly 10% of the primary energy supply, global biofuel production grew to over 40 billion gallons in 2018, primarily in Brazil, China, and the

United States. This production represents about 3.7% of the transportation fuel demand.[209]

American automobiles currently run primarily on fuel that is 90% gasoline and 10% corn ethanol. Luckily, or should we say by the power of the market, America produces 1.02 million barrels of ethanol a day which matches up at a 90/10 ratio with the 9.3 million barrels of daily dispensed gasoline. Unfortunately, we consume 40% of the corn crop to produce 10% of our fuel at the cost of a 5% increase in corn prices. Further, taxpayers spend more than $4 billion per year in tax credits, 71% of all farm tax credits, given to farmers to raise those crops.[210] Of the 14 billion bushels of corn produced in 2018, 5.5 billion made biofuel, 0.5 to create high-fructose corn syrup, and just 0.2 billion bushels to create food and cereals. Biofuels are the largest consumer of corn, followed by animal feed. So next time you drive through the cornfields of Iowa, don't think food, think animal feed and automotive fuel.

One may argue that we won't need biofuels as the plan is to electrify our vehicles. While the average age of an automobile in America is 11.5 years, the average age for farm tractors is over 20 years, and the tractors in the most demand are 30-40 years old.[211] Increasingly farmers are becoming more attached to older farm equipment as being more practical and not having all the unnecessary luxuries. Items such as farm equipment, factory vehicles, and generators will continue to operate for decades. Thus, fuel alternatives will be necessary.

Since corn is a crop, it takes fuel for the tractors and combines, water for the plants and land. A lot of land, some 31 million acres, 48,500 square miles. Enough crop area to cover Rhode Island, Delaware, Connecticut, New Jersey, Massachusetts, New Hampshire, Vermont, and Maryland. All this land for just 10% of our fuel, want more fuel? It takes even more land! And as we will see, needing more land will become a reoccurring theme. If the U.S. were to replace all fuels with biofuels from corn, realizing that the U.S. gets twice the corn per acre than the rest of the world, the land required would be 20% more than the U.S. landmass. It is unfortunate that so much effort is spent to create the lowest quality fuel currently available due to its low energy density. In the end, the problem with all the renewable sources of energy is that they simply do not have enough energy density, enough concentrated oomph to do the job.

As Vaclav Smil writes, "The fact that wind, solar, and biomass have incredibly low energy density per square meter means that a fully renewable system to replace the 320 GW of fossil-fueled electricity generation and 1.8 TW of coal, oil, and gas with biofuels would take up to 50% of America's territory, 1.81 million square miles (250-470 Mega hectares) since the average power density of biomass is just 0.45 W/m² to produce liquid biofuels."[212] And this was stated in 2015 when America's electricity demand was less than half of what it is today.

The direct impact of any energy solution is determined by the energy density of the solution. As Michael Shellenberger points out, "As such, coal is good when it replaces wood and bad when it replaces natural gas or nuclear. Natural gas is good when it replaces coal and bad when it replaces uranium."[213]

Renewable energy is often like recycling; while it is excellent in theory, it is more difficult in reality, where only 5-10% of what goes into a recycling bin gets recycled. Cities and individuals are firmly in favor of recycling but do a poor job of collecting and sorting their recyclable material and further, even if collected as recyclable, a large percentage ends up in the landfill due to lack of technology to use the material, the lack of a market or the cost being such that new material is cheaper than using recycled material. Being all for an idea does not make it practical, profitable, or make sense as a plan for the future.

The **Green Solution's** goal is to use these problematic renewables to provide all our power on the electrical grid. Knowing what you now know, is this a good idea?

9

Can We Electrify Everything?

Globally, on an average day over the last 11 years, 314,770 people got access to electricity for the first time. The total number of people without electricity, most of whom reside in Sub-Saharan Africa, fell below one billion for the first time in 2015.

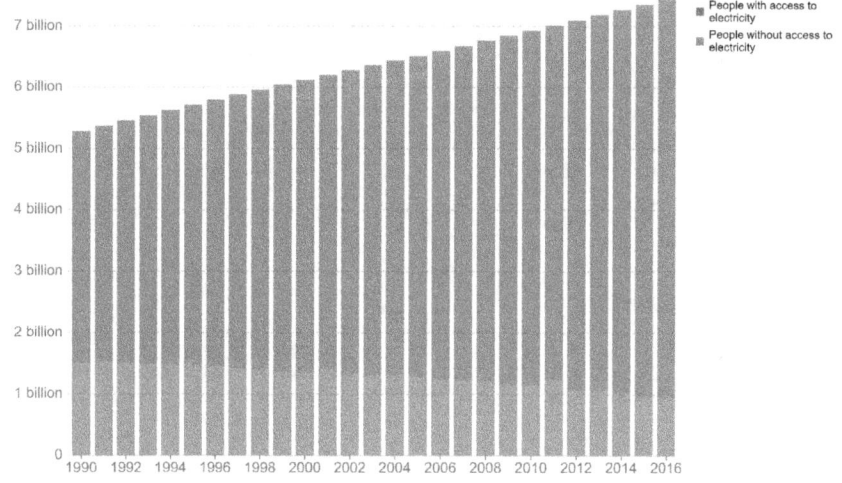

Figure 50 Number of People with and without electricity access

My friend, Cedrick, from Cameroon, describes that sub-Saharan Africa has a two-tier electrical system. Those who live within the cities have electricity while those who live in the suburbs and outwards don't. Although those who live in the city have homes which are wired for power, that does not mean having access to electricity, as it is off for days up to a week, without notice. To those in the rural areas, his company sells solar-powered lighting and television systems. To those who have never had electricity, access to a handful of LED lights on a single string is a miracle, something no longer appreciated within the developed nations.

There are nearly 3,000 individual electric distribution systems across America, each with its own set of challenges between weather, terrain, demand curves, and utility regulations. On each of those systems, customers expect power to be continuously available and it would be within the specified voltage and frequency ranges. It is estimated that the average electrical customer attempts to contact the power provider within 15 minutes of a power outage. In fact, the flood of incoming phone calls became so crippling to their phone systems that it was worthwhile to create methods to text individual customers to inform them of repair status. To return electrical service after significant storms, linemen now rush in behind the storm to repair lines, sometimes from hundreds of miles away. To Americans, the lack of power is viewed as a disaster. The average electrical consumer is without power for 8 hours a year, less than 2 hours outside of significant weather events. Among the states, North Carolina performs the worst at nearly 29 outage hours, while the Dakotas are best at around 2 hours per year. The world's average is 141 hours per year, with some remote villages still experiencing 4 to 8 hours of outage daily. A 2018 report by the World Bank expressed that for some remote Indian villages, every 1-hour increase in daily power outages reduced the average household income by 0.5%.[214] For these villages, outages aren't a disaster, just an inconvenience. The fact that Americans are so concerned with their few hours of outage per year, while elsewhere in the world, it is daily outages, should really be a bigger concern.

As America undertakes to convert its plentiful, reliable, and cheap energy grid to one that would be governed under the principles of the **Green Solution**, we must consider the implications of that change. Under the new grid design, we must ask two questions:

1) Can 90 to 100% of that electricity be renewable?
2) Can we electrify everything?

The answer to either question is yes, but the answer to doing both is probably not. And if we do either, will the grid still be the reliable, plentiful, and cheap grid we have come to rely upon? A saying attributed to Confucius says, "A man who chases two rabbits, catches neither."

The mainstream of discussion in America today divides electrical power sources between renewables and non-renewables. All the sources of energy yet devised by man come from nature. The issue is how we divide that energy between those whose energy source is today's nature and those from the nature of the past. The flowing rivers, shining sun, blowing wind and live plants against the long past effect of the sun on plants of another era and elements generated in the heat of a supernova.

To divide the sources this way is simplistic. Some better ways to evaluate energy generation methods would be to look at them as independent, reliable, controllable, sustainable versus interdependent, unreliable, uncontrollable, unsustainable. These four new ways of looking at power sources reflect how the energy is being used and the time frame for which they are being evaluated. Controllability reflects a source's performance over a few minutes, and reliability considers the performance over seasons, while sustainability evaluates a power source over lifetimes.

Table 9 Evaluations of Power Sources

Source	Renewable?	Independent	Controllable?	Reliable?	Sustainable?
Coal	No	Yes	Yes	Yes	Yes
Oil	No	Yes	Yes	Yes	Yes
Nat. Gas	No	Yes	Yes	Yes	Yes
Solar	Yes	No	No	No	No
Wind	Yes	No	No	No	No
Hydro	Yes	Sometimes	Yes	Yes	Yes
Nuclear	No	Yes	Yes	Yes	No

When considered from the outside, the electrical grid as a whole can seem almost magical. For within any grid, there is no manner in which excess electricity can be saved for later or excess demand selectively ignored, and yet it is magically able to continuously balance electrical generation with the ever-changing need. It must maintain this balance, as an out of balance grid may risk damaging millions of dollars of power generation equipment. The operator's requirement to maintain this balance places certain limits on the behavior of the energy sources that the grid utilizes. If the energy types at the operator's hands don't meet certain criteria, the grid may fail to properly maintain this balance. We know that this balance is successfully maintained because we have that average of just 8 hours of outage a year.

Independent / Interdependent

Basically, fueled plants are independent, and renewables are interdependent.

Electricity demand can be viewed as being made up of two distinct parts. The first part is the baseload, which is the part of the demand load which is always present, regardless of weather or time of day. It can be viewed as the minimum demand over 24 hours, roughly 32,000 Megawatts for the Texas day in Figure 51. The second part is the variable load, the daily fluctuations as demand increases in the afternoons, about 19,000 Megawatts in our example day.

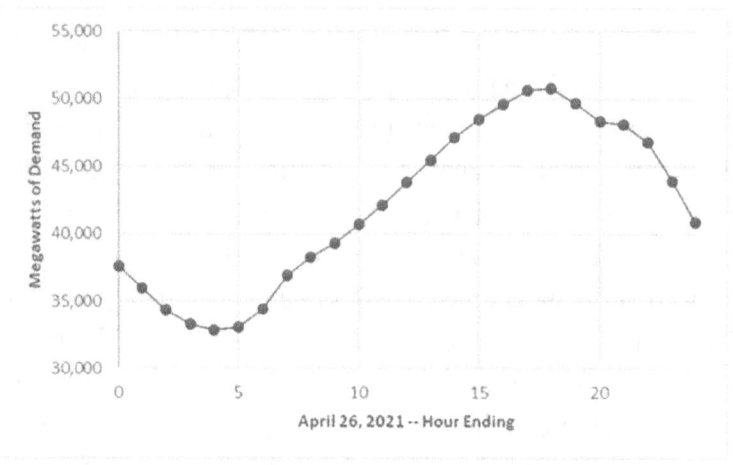

Figure 51 Texas (ERCOT) Demand Load for April 26, 2021

There is one further load of electricity generation that must be considered, that being peak demand. The peak demand represents the extreme situations beyond the usual generation conditions, typically due to a hot summer day or a frigid winter night.

Within America, the electrical grid is divided into many smaller grids that can be operated in one of two ways: a public utility or as a Regional Transmission Organization (RTO), or more commonly understood by the population as regulated and deregulated grids. The "deregulated" grids are run by the RTOs, such as ERCOT for the Texas grid, and are heavily regulated, regardless of the nomenclature. On an ongoing basis, within any grid managed as an RTO, power generators, those who operate the power plants, make bids and receive payment in exchange for promising to be available to generate power when requested. When the grid operator asks that a source provide the baseload, it needs to be able to do just that, without issues, and are usually provided by those slow to respond but steady sources such as nuclear or coal. When a power source supplies the baseload, the source is independent if there is no need for a backup source.

The **Green Solution** proposes handling the baseload using renewables like wind and solar. However, if the weather doesn't cooperate, they cannot provide the baseload, and therefore they require a backup. Usually, this is a natural gas-powered facility. As Robert F. Kennedy Jr states in his role as an environmental activist, "The plants that we're building, the wind plants and solar plants, are gas plants." Studies have shown that for every 1 GW of interdependent renewable sources placed on the grid, 1.14GW of independent sources are required for backup in anticipation of when the weather isn't favorable.[215]

The interdependent sources need a backup because clouds pass overhead, and the wind starts and stops as it pleases. Unfortunately, it is even more troubling as the clouds and wind changes happen at a much faster rate than gas-powered plants can respond unless they are idling in case of need. Thus, the backups are burning fuel and consuming energy just in case they are needed. Applying these interdependent sources to variable or peak loads is no more helpful since it is unpredictable when the backup is required. Moving to an all-renewables grid will require that the regional grids be interconnected with the proposition that wind will always be blowing somewhere; therefore, if the wind is not generating power in your state, it will always be blowing somewhere. Thus, even the Green Solution plans recognize that renewables are interdependent, and the only question is what is their backup.

Controllable / Uncontrollable

Within the peak-load, when the demand for electricity goes up, at any given time, the grid operator will ask power sources to deliver promised power. To do this, an energy source must be controllable. The operator of the generator needs to figuratively have a knob to dial up or down to supply more or less power. These adjustments are, of course, managed by software within the grid. This prompt response is easiest accomplished with gas- or oil-fueled facilities. Feed more gas or oil into the system, and you get more power. Less fuel gives less output. Facilities like hydro, coal, and nuclear are very much controllable but take significant amounts of time to change their rates. So the current grid utilizes gas peaker plants, facilities that use natural gas for the express purpose of generating quick, on the spot power as needed.

Unfortunately, when it comes to some of the renewables, they aren't controllable. When the grid operator asks for more power, and there is no shining sun, then the solar units cannot provide power. Likewise, if the wind is not blowing, the wind generators cannot help to meet the increased power demand. Grid operators need to be able to plan power generation at least a day in advance, and even with accurate weather forecasting, this is difficult when dealing with uncontrollable sources. When a source isn't controllable, a controllable backup source is again required, so once again, those backup gas plants are needed.

Reliable / Unreliable

It has been shown that certain renewables, namely wind and solar, are neither independent nor controllable, which means that the grid operators would struggle to integrate them into their grids.

The reliability of a source is a determination of its independence of outside factors which could influence its ability to deliver power upon request. Hydroelectric dams are highly dependent on the level of the reservoirs of water that lie behind them. Droughts may have the effect of lowering the amount of water available to push through the dam. While environmental and economic issues may restrict the amount of extra water that can be held. While hydro is controllable and independent, it is only somewhat reliable as it is affected by the water levels.

Likewise, solar and wind suffer from seasonal variations. Solar plants in the northern states will be affected by low sun angles in the winter. Just as wind facilities will be affected by the lower average wind speeds in the summertime. This means that these sources are available during some times of the year and less so at other times. Gas, coal, and nuclear plants provide power consistently regardless of the weather, time of day, or season and thus, a grid operator can rely upon them being there when power is requested.

Electrify Everything?

The twin goals of electrifying everything have been tried before, and there are lessons that we have already learned. Much is made of Germany's effort to move to 100% renewable energy by 2050 (updated from the original goal of 2025) and the fact that they achieved 100% renewable energy for a moment in time at 6:00am on Jan 1, 2018. Headlines around the world noted the achievement. On a particularly windy day, the accomplishment was achieved, a day on which most people stayed in bed after a late New Year's Eve celebration. Still, as the renewables reached a generation level that satisfied 85% of the demand, the shutdown of the non-renewables made the moment possible.[216] Still, after spending $580 billion on the project, renewables annually provide 46% of Germany's electricity, while 40% remain supplied by fossil fuels. The two issues with Germany moving forward in the production of more renewable energy are: first, the wind is generated offshore in the Baltic Sea while the largest demand is in the southern part of the country, and secondly, the grid is not able to handle the shipping of power from north to south. So, much of the northern power is sold to countries to the north (Denmark, Poland, the Netherlands) while the south purchases power from other countries.[217] At times, Germany has paid other countries to turn off their power sources so that Germany can do something with all that wind power they generate. So why is

Germany so far ahead on the renewable game? The answer lies in the Renewable Energy Act surcharge (6.24-euro cents per kilowatt-hour)[218], which all power customers pay, which along with taxes, can be as much as 50% of the household bill.[219] Once all surcharges and costs are figured in, the "green energy producers were paid on average 14 times more for their electricity than it was worth at actual market prices."[220] If renewable electricity is so expensive, why does the renewable industry advertise itself as cheaper than the current technologies? Simple, they add to the cost of fossil fuel electricity the expected cost of $1.2 trillion in added health and environmental damages, including 85,000 premature deaths through 2050. All those dollars and deaths are the result of modeling, and no actual costs or deaths are attributed.

Even with an experience where it hasn't worked as planned, there are still two distinct, fully fleshed out plans to implement the fossil fuel free world: the Solution Project, which plans for 100% clean and renewable, and the 2035 Report, which plans for 90%. As the other plans, such as the Biden Unity plan and the Green New Deal, do not reflect thought-out complete plans, we will use the Solutions Project and 2035 plans as substitutes.

Grid Level Storage

To electrify everything using uncontrollable sources, some form of energy storage will absolutely be required. In the new green environment, energy storage will be necessary so that energy can be released during two particular scenarios: when uncontrollable power is not available and during periods of excessive peak demand. Further, storage should allow a grid system to require less overall generation capacity since the demand can be smoothed out over the course of the day.

The goal of any energy storage is to convert that which is temporary, i.e., electrical energy from the grid, to that which can be contained and controlled. The storage duration may vary in length depending on the storage methods; some methods may be stable for extended periods (days, months, years) while others may be more temporary (minutes, hours, days). Energy storage can exist in many forms: electrical, chemical, and mechanical.

	Max Power Rating (MW)	Discharge time	*Max cycles* or lifetime	Energy density (watt-hour per liter)	Efficiency
Pumped hydro	3,000	4h – 16h	30 – 60 years	0.2 – 2	70 – 85%
Compressed air	1,000	2h – 30h	20 – 40 years	2 – 6	40 – 70%
Molten salt (thermal)	150	hours	30 years	70 – 210	80 – 90%
Li-ion battery	100	1 min – 8h	1,000 – 10,000	200 – 400	85 – 95%
Lead-acid battery	100	1 min – 8h	6 – 40 years	50 – 80	80 – 90%
Flow battery	100	hours	12,000 – 14,000	20 – 70	60 – 85%
Hydrogen	100	mins – week	5 – 30 years	600 (at 200bar)	25 – 45%
Flywheel	20	secs - mins	20,000 – 100,000	20 – 80	70 – 95%

source: The World Energy Council

Table 10 Characteristics of selected energy storage systems

Typical mechanical storage proposals are accomplished utilizing pumping water, compressing air, or spinning flywheels. While the electrical means represent the more commonly thought of methods, including batteries or through creating of hydrogen gas. Each method has its benefits and issues. Most mechanical storage methods work at an industrial scale and have longer lifetimes but suffer from the need to convert the energy back into a useable form after storage. Pumping water behind a hydroelectric dam does store energy but does nothing to increase the amount of energy available to the grid, as the maximum output of the dam turbines is unchanged.

Most storage applications currently planned have limited timeframes in mind. Currently, several companies promote the use of solar + storage to homeowners and corporations to reduce their energy expenses. However, these storage systems are designed to be short-term solutions, charge in the morning when the energy is cheaper, and use the power in the afternoon during the more expensive peak energy demand. The goal of these systems is to even out the energy demand over the day, while also providing power during a passing cloud. This short-term storage is required as demand peaks in the afternoons when the population returns home, just as the solar power fades away—unfortunately, solar power peaks at midday which isn't high on the demand curve. There are no plans for any storage systems to handle the weekly, monthly, and seasonal shifts in power. The type of storage that would be required if, for instance, solar energy is collected in the summer and storage is used in the winter.

The International Energy Association (IEA) estimates that the world will need 943 Gigawatts of storage by 2040. Consider that as of 2018, the U.S. energy grid had 0.71 Gigawatts of battery storage capacity available. Of all storage on the U.S. grid, 95% is in the form of pump-storage hydropower (PSH), consisting of around 22 Gigawatts of energy. Basically, PSH represents pumping water that has been through the hydroelectric dam back into the reservoir so that it can flow through again. Obviously, there is energy lost in the act of pumping the water, but if there is excess energy to be had, it is a good use. However, it is limited to where such hydroelectric dams exist and only so much as to not flood the other services provided by the reservoir.

Building storage from lithium-ion batteries runs into issues with not having enough lithium. A single Kilowatt-hour of storage requires 77 grams of lithium metal. Enough storage for one hour of grid storage would take 122,000 tons of the metal, about 3 times the global production in 2016. If dedicated strictly to grid storage within the United States, the annual production of lithium would provide just 18 minutes of backup. Not counting all the other uses for batteries made from the material. Add to this limitation the recent election of "radical blocs" to a majority position within the Chilean government, a country that is the world's largest lithium producer. The new government campaigned on "economic nationalism," insisting that any license to any lithium mining should include building the batteries there, with some going as far as demanding the manufacture of the entire electric vehicle.[221] The mining operations are being targeted for renationalization, else their "old, grandfathered, licenses" might be revoked. All that we can be certain of is that "they're going to change the basic rules of the country."

That estimated electrical storage requirements will be vastly insufficient since we are subjecting the electrical grid to two significant changes, 1) Move everything to renewables and 2) electrify everything.

Can the grid be 90 to 100% renewable?

The best place to start when considering the grid is asking how much electricity we now use. From 1950 to 2005, electrical demand grew at a steady 2.6% per year. The Energy Information Agency (EIA), a division of the U.S. Department of Energy, is tasked with collecting past and current statistical data relating to energy, data which this book relies heavily upon, and issuing projections about the future use of energy. In 2005, EIA issued an extensive report on energy usage over the next 20 years, out to 2025.[222] The report indicated that electricity usage would grow from the current 2005 value of 3,653 Terawatt-hours to 4,480 in 2020. The actual 2020 electrical consumption came in at 3,802 Terawatt-hours, an error of 16%. In fact, rather than demand continuing to grow, it leveled off since then, being steady at around that 4,000 Terawatt-hours. The electricity demand has leveled as electronic efficiency improvements (flat-screen TVs, laptops, cell phones, DVD) have matched the rate at which we have created new devices to plug into the grid. In fact, the latest forecasts have only modest growth due to a growing population between now and 2050. The future growth from our current 4,000 up to the projected 4,500 Terawatt-hours is largely driven by the three largest drivers of electrical demand over the next couple of decades which are expected to be business as usual growth in electric vehicles, data centers, and marijuana cultivation.

All total, by the end of 2020, the United States had about 1200 Gigawatts of electrical generation capacity, of which 43% natural gas, 20% coal, 16% wind and solar, 9% hydroelectric, and 9% nuclear, with a few other minor sources. With 1200 Gigawatts of generation capacity, there is a potential to generate 10,512 Terawatt-hours of electricity annually if all the sources ran at 100% every day of the year.[5] Roughly, America uses 38% of its potential electricity generation capacity. This may seem a bit of a waste, but all that capacity is needed on those extreme weather days of deep winter or scorching summer, typically a peak demand of 770 GW. While 770GW still represents less than the full grid, unfortunately, the peak demand isn't always where the sources are located. Any renewable grid must replace those 4,000 Terawatt-hours with enough margin (currently 150%) to absorb the extreme events.

The renewable grid program is often referred to by the acronym WWS (Wind, Water, and Sun), consisting of wind turbines, hydroelectric plants, solar panels, and utility-sized solar installations. Any WWS program would need to replace all the non-renewable sources with one of these. Several proposals are serious in their intent of implementing an actual WWS grid, each with a rather complicated scheme by which states use different mixes of energy generation sources based on their geography. As such, it isn't easy to describe the system as a whole.

---

[5] Potential Generation = Capacity in GW * 8760 hours / 1000 (to convert units)

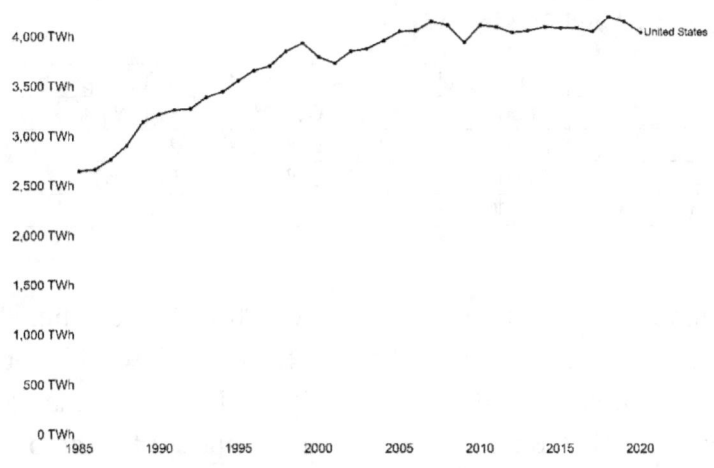

Figure 52 United States Electricity Consumption

Both the 90% renewable 2035 Report from the Goldman School of Public Policy at the University of California – Berkeley and the 100% renewable Solutions Project by Stanford engineering professor Mark Z. Jacobson call for 1500 GW of new wind and solar capacity to be installed over the next 15 to 25 years to replace the existing fossil fuel powered electrical grid. Each differs in the combination of the sources which they would implement. The 2035 Report is somewhat vague, but with Jacobson, the documentation is specific, and websites[6][7] are available to look up the plan for each state.

Renewables currently provide 25% of the generation capacity and deliver around 13% of the generated electrical power. To date, America has installed slightly more wind than solar, 121 GW versus 97 GW, but solar has been gaining ground in recent years. In the past year, the United States installed 14.2 GW of wind and 19.2 GW of solar generation, nearly 80% of which was built in China.

---

[6] thesolutionsproject.org
[7] https://www.nationalgeographic.com/climate-change/carbon-free-power-grid/index.html

Table 11 WWS Plan Build Outs by End of Plan

State of Plan	Wind Turbines	Wind Capacity	Solar Capacity
Current	69,000	121 GW	97 GW
New Green Deal	130,000	250 GW	200 GW
2035 Plan	300,000	596 GW	515 GW
Jacobson Plan	496,000	715 GW	620 GW

The WWS plans call for the building of 90-100 GW of combined power for the next 15-25 years. Any such proposal would find specifications difficult as technology will undoubtedly change considerably over that time frame, so some proposed technology may not yet exist.

The basic Jacobson plan calls for:

> • A half-million giant 5-MW wind turbines (not currently available on the market) on acreage equal to New York state, Pennsylvania, Vermont, and New Hampshire, and in open sea regions equal to West Virginia
> • Billions of solar panels on 75 million homes and nearly 3 million business using on land equivalent to Maryland and Rhode Island
> • Concentrated Solar Power (future technology) on land equivalent to Connecticut
> • Building 4% extra capacity for peak demands[223]

The projected business-as-usual approach, meaning letting the market decide, is expected to install an additional 36.7 gigawatts of new generation capacity by 2030 at the cost of $48 billion. The price for implementing the 100% renewable program would be similar to "the nation's most ambitious endeavors: the Apollo program, the interstate highway system, the nuclear bomb, and the military's World War II arsenal. This transformation would cost roughly $15 trillion, or $47,000 for each American, for building and installing systems that produce and store renewable energy."[224] Roughly $1

Trillion per year in spending, although we will soon find that that won't be enough.

Table 12 Annual Costs of Suggested Equivalents

	Cost in 2020 dollars	Annual Cost
Apollo	$215 Billion	$16.5 Billion
Interstate Highway	$615 Billion	$30.7 Billion
Manhattan Project	$23 Billion	$7.7 Billion
World War II arsenal	$2 Trillion	$500 Billion

Within sailboat racing, the primary tactic used to put distance between yourself and another is to position your boat such that the wind shadow from your sail, that area behind it where the wind is disturbed, upon the other guy's sail. This keeps their sail from working effectively and slows them down. The same effect is true for wind turbines, they leave wind shadows, large ones, three times the turbine's height. So the turbines can only be placed so close to each other. While the Jacobson plan proposes to use 1% of the land area, the National Renewable Energy Laboratory's estimate, based on real-world experience, increases that size by 4x, upwards of 262,000 square miles, 4% of the nation's landmass. The project has carefully found useable locations for each of the 496,000 proposed turbines, a number of turbines based on calculations of how much power they would need to generate and shoehorned them into the geographical areas where wind power makes sense. The NREL's estimate of spacing means that ¾ of the turbines will necessarily be situated on land that is less productive due to geography.

The solar aspect of the Solution project plan calls for solar to generate 46% of the nation's electricity, 11% from the rooftops of 75 million homes, 15% from the rooftops of commercial and governmental buildings, and 20% from utility-sized solar plants. Current solar installations, although 80% of the capacity of wind, generate just ¼ of the electricity of the wind plants, contributing just 2.3% of the 2020 electricity generation from all sources. The plan calls for 20% of all power and the entire 4% overbuild to come from concentrated solar power (CSP), which uses mirrors to heat water to run through boilers. To date, the U.S. CSP capacity is 1.815 GW, basically the output of a single average wind turbine. As we will see in Chapter 11, so few CSP facilities are in place because local residents fight their placement due to the dangers to the local ecology. These facilities take 5 to 10 acres per megawatt of generation capacity. The result is a call to take up another 0.9% of the landmass, all of which must be in the American southwest.

America currently has a large overbuild of capacity on the premise that each region should be self-sufficient for peak demands. The proposed renewable grid assumes that this autonomous nature is unnecessary and, in fact, undesirable. With the thousands of miles of underground high-voltage power lines proposed by President Biden's Infrastructure Bill, all the power regions can be interconnected, and the shortfall from any one area can be made up by another area with surplus power. With the uncontrollable and interdependent nature of renewables, this becomes a desired outcome.

Is the number of turbines and solar facilities sufficient? Overall, the plan uses optimistic numbers for capacity factors based on promises of future technology. To make the goal, the system build must start soon, using current technology, and the rate of construction will necessarily begin at 2-3x the current rate and then grow exponentially from there. The Green New Deal and President Biden commit to American-built devices, currently, 80% of which are of Chinese manufacture, so many more times increase in manufacturing will be required. Limited to 75 million of the 108 million single-family homes and limited useful wind generation locations, the plan ends with a final requirement, a 40% reduction in electricity demand. In the past history, even with efficiency improvements, the only reductions in electrical demand have been during recessions, with their accompanying decrease in industrial output.

The metabolic rate, the rate at which a nation burns energy to produce and flourish, will need to be cut by half or more. Back to the metabolic rate that the nation had before the industrial revolution. Ultimately, the grid is designed to be unreliable, scarce, and more expensive.

Can we Electrify Everything?

The next question is how much electricity will be needed if "electrify everything" is implemented. Taking this step requires moving residential and commercial buildings, industry, and transportation from the secondary fuel side of the great divide, discussed in Chapter 4, to the electricity side.

Decisions of politicians and market forces will determine the rate at which electrification technologies are adopted, affecting just how quickly those items will move from one side of the divide to the other. The pressure from the left will encourage politicians to take action, preferably before the next election, which means that aggressive plans for political reasons may overrule the practicality of the ideas.

Of the total energy usage in America, 28% is for transportation, 23% for industrial, 7% for residential, and 5% for commercial. The remaining 38% of the energy goes into electricity generation, 13% comes out as sold power, and 25% is used to generate the electricity and is lost. So, of the 101 Quadrillion BTUs of energy, roughly 62% of all energy would need to move across the divide to "electrify everything."

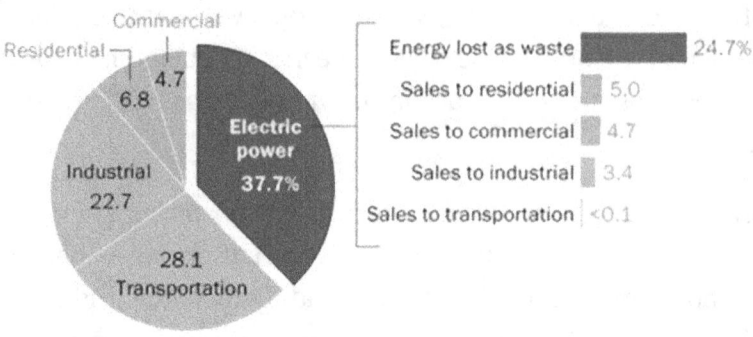

Figure 53 Energy Consumption by Sector, 2018

There are currently an estimated 282 million registered vehicles in the United States. After 20 years on the market, all-electric and hybrid vehicles have a 2.08% share of the vehicles in operation. In 2019, 4.7 million vehicles were sold, of which 4% were hybrid models, and 2% were plug-in hybrid or fully electric models, nearly all Teslas. The 2035 Plan on electrification of transportation uses "a scenario in which EVs constitute 100% of U.S. [light duty vehicles] sales by 2030 and 100% of [medium duty vehicles] and [heavy duty trucks] sales by 2035, while the grid reaches 90% clean electricity by 2035 and substantial charging infrastructure is deployed."[225] They expect to encourage the sales by providing incentives, providing charging stations, and limitations to the market choices, noting Volvo's commitment to sell 100% EVs by 2030. Sales will necessarily go electric when the showrooms are limited to those models. The report indicates that the number of required charging stations will be around 72 million privately owned and 9 million publicly accessed, based on 75% of vehicles being charged at home. Light Duty vehicles (cars and light trucks), which account for 90% of motor vehicle traffic, are expected to use between 570 and 1140 Terawatt-hours by 2050.

Heavy Duty vehicles, trucks and transports, account for 4% of the vehicles in operation but use 20% of the transportation fuels. Converting these to electric will require the challenging goal of electrifying every mile of the major highways to install 900,000 charge stations for the medium and heavy-duty truck market spread across 2,700 truck stops.

The study shows that electrification of vehicles raises the peak load from the typical maximum of 770GW to 1050-1250GW. This does, however, assume that vehicle recharging will be at home at night in a manageable manner.

The greatest fossil fuel usage in the residential and commercial spaces is for heating. With 60% of home heating and 35% of cooking dependent on fossil fuels, these homes will need to be converted to heat pumps and electric stovetops. In 2019, the U.S. natural gas demand set records at 145 billion cubic feet per day (Bcf/d), with single days exceeding 150 Bcf/d. During winter storm Uri, the natural gas system delivered about 80 Bcf/d to homes and businesses. For the electrify everything grid to deliver the equivalent power in the form of electricity would require an additional 1200GW of capacity, roughly as large as the existing grid.

The general idea of heating our homes with electricity through heat pumps runs into a flaw. The typical heat pump loses efficiency between 25° to 40°F, depending on design, and often fails to work at all around -4°F. I can personally attest to this shutdown, having lived through winter while owning a heat pump in Allentown, PA. Therefore, the WWS plans have no intention of moving home and business heating to the electrical grid, rather they plan to use Underground Thermal Energy Storage (UTES) in the form of borehole energy storage (BHES). It is unlikely that the reader is familiar with these technologies, which will be heating homes everywhere soon. In this method, water is pumped through roof-mounted solar heaters, and then the water is sent through a network of underground pipes so that the surrounding soil retains the heat. This heat can later be retrieved and, through a system of radiators, heat the living space. While this may work in some regions of moderate weather like California, it is likely to be as popular in Alaska, the Northern Plains, or New England as those heat pumps which shut down when they are needed most. Admittedly, they have forgotten that the space for those water heaters has already been taken up by solar panels for electricity generation. All those water pumps will need that electricity to operate.

The current published estimates project that by 2050, the electrify everything grid would require 7,500 Terawatt-hours of generation to cover the 4,500 Terawatt-hours for our current electrical usage and additional 3,000 Terawatt-hours for moving items to the electrical side of the great divide. However, if the BHES heating systems don't work, the demand will be well over 9,500 Terawatt-hours.

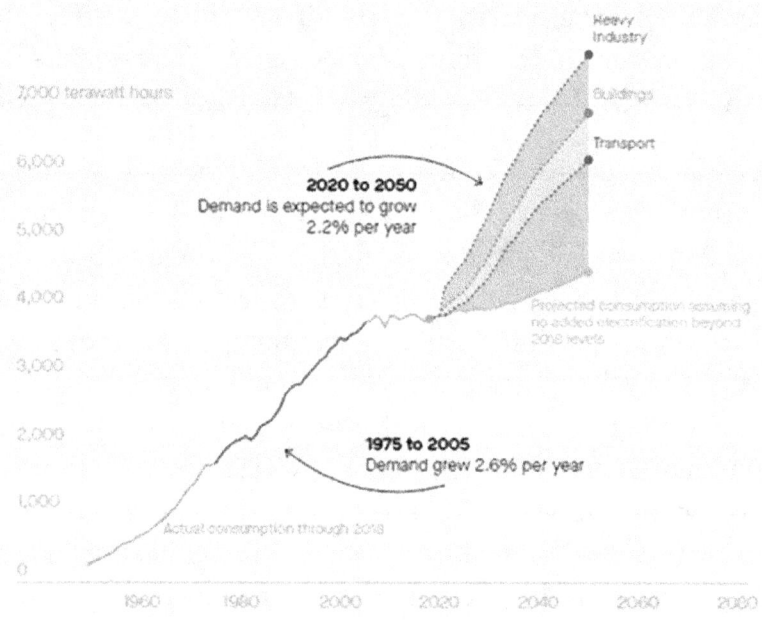

Figure 54 The Demand on the Electrify Everything Grid

Before introducing the new demands onto the electrical grid, the renewable grid projects called for installing 90-100 GW of WWS capacity each year. Once everything is electrified, that will need to increase by roughly 30% according to their numbers to 120-130 GW of new sources each year and remember that 35GW is our one-year maximum installation to date.

Increasing the need by 30% means we now need 645,000 wind turbines and 97.5 million solar homes, at a 30% increase in the $15 trillion cost. Our current grid runs at 38% of capacity but the electrify everything grid would be running at 72% of capacity, which will only work if intermittent routine outages become acceptable. It is unlikely that any system already running at 72% of capacity can handle extreme events like summer heat or winter freezes. At a minimum, it would be necessary to shut down some demand during these events, whether industry or vehicle recharging. So, more realistically, even more generators will be needed.

Sustainable / Unsustainable

A power source's sustainability depends on whether additional and replacement units can be built, fuel supplied, and maintenance maintained. The fully renewable grid will take some 15 to 25 years to install. With the typical 15-20 year lifecycle of many of the components of the WWS grid, the earliest elements will be reaching the end of life as the last structures are put into place.

With a wind turbine, the physical structure has a lifetime of 60 years, but the blades, the most expensive element, need replacing every 10 years and the mechanical internals to the nacelle need major overhauls every 20 years. These upgrades will represent 10% of the initial costs. This, of course, means that the overhaul schedule will begin long before the last wind turbines are built. With the required 645,000 turbines, 100 wind turbines must complete overhaul daily. Although, without petroleum being processed, it is unclear today where the materials for the maintenance will come from.

The somewhat longer lifetime and the lack of moving parts within solar panels mean that panel replacement can wait until the last of the necessary panels are built. Still, once the panels are produced, the daily replacement rate will be 17 million square feet to keep the grid at full operation.

In 2020, the United States consumed 3.2 million tons of copper, of which one-third was imported. The current state of the art in design means that solar panels need 5 tons of copper per MW of capacity while turbines require 3 tons. This requires 31.7 million tons of copper per year at the suggested build rate, plus the copper needed for power lines and interconnection hardware. The United States alone would require one-third of the world's supply, shorting all other needs for the metal. Additionally, the Solar Panel's 13 mg / W Silver requirements would take up 90% of the proven silver reserves. Each would stress the production of all the other devices in our lives when the material would need to be dedicated for electricity generation.

At the beginning of the chapter, two questions were asked, could the grid be fully renewable and whether everything could be electrified. Further, if those things could be accomplished, could it be achieved with the characteristics of today's grid of being reliable, plentiful, and cheap.

Determining the cost of electricity in a renewable grid would be impossible as there are too many variables. An analysis by MIT produced by Jesse Jenkins, estimates that a 100% decarbonized renewable electrical grid would cost between $150-300 per megawatt-hour (current costs average $105 MWhr). Thus the average American can expect an increase in their electrical bill of between 43 and 286 percent, as much as an extra $3900 per residence per year.[226] Although, admittedly, that is before the 50-100% increase in the kilowatt-hours used when electrifying everything is completed.

Can a 100% renewable, electrify everything grid be built? The answer is yes, but it will take more land and resources than we can reasonably use, it will cost significantly more, and it will likely be highly intermittent. As Mark Twain noted, "land, they aren't making it anymore." Land may well become the ultimate limiting resource. Without enough land, ultimately there won't be enough electricity. The implications of intermittent electrical power include:

- Loss of perishable foods and medications
- Loss of water and wastewater distribution systems
- Loss of heating/air conditioning and electrical lighting systems
- Loss of computer, telephone, and communications systems, including emergency services and the internet
- Loss of public transportation systems
- Loss of fuel distribution systems and fuel pipelines
- Loss of all electrical systems that do not have back-up power.[227]

The Big Lesson

A House of Cards / Jenga Tower / A Fall of Dominos

There is a single lesson for which this book was written. If the reader understands this one lesson, they will understand the issue. The task is to see how the single simple statement or proposal of today's activist or politician has far reaching consequences on how we will all live our lives.

Today, the energy world is essentially represented by two items, the barrel of oil and the electrical outlet. That barrel of oil consists of almost half gasoline. Another quarter is other fuels like diesel and jet fuel. The remaining quarter is those compounds used for petrochemicals and things like asphalt. The electrical outlet represents the 50% natural gas-powered electrical grid that is reliable, plentiful, and cheap, which with rare exception, typically 8 hours a year, always there when we need it.

Into this world steps President Biden with several executive orders and proposals: an all-electric federal vehicle fleet, federally installed electrical recharging stations, and regulations to encourage the purchase of more electric vehicles.

Now the dominos begin to fall. When enough electric vehicles replace their gas-burning counterparts, the demand for gasoline goes down, resulting in falling gasoline prices. Eventually, this limits the profitability of gasoline, which, remember, makes up nearly half of that oil barrel. While at the same time increasing the demand on the electrical grid, as the power to recharge those vehicles comes from somewhere. As oil producers cut back on producing unprofitable gasoline and vehicle manufacturers build more electric models, consumers will see the electric vehicle as a better choice, pushing more sales away from gas-powered towards electric. A death spiral for gas engines and the gasoline industry begins. All by design, of course, this was the goal of the President's proposals in the first place.

The second step occurs when the President mandates no more fossil fuel powered plants while closing nuclear and cutting back on hydroelectric. So additional capacity must be made by renewables, which are uncontrollable, interdependent, and unreliable. As demand increases due to all the new vehicles, more renewables are brought onto the grid, requiring more land to be used for the placement of the wind turbines and solar panels. Those vehicles can now be recharged, well, at least when the sun shines and the wind blows.

So what happens with the rest of the barrel of oil? Without profits from gasoline, soon processing the whole barrel is unprofitable, so no diesel or aviation fuel. Unless they find they can profit by raising the prices of the other fuels, which has two implications, costs skyrocket for everything shipped by truck or plane, which is everything, and ½ of the barrel of oil is now useless waste. And what about all those petrochemicals upon which our lives depend?

As the oil industry shuts down refining, the older personal vehicles and farm equipment must run off corn-based ethanol, so more land must be reserved to produce biofuels. While at the same time, the reduction in petrochemicals means that more feedstocks must be generated from corn-based ethanol. The feedstocks that are required to make our medications, clothing, technology, and the wind turbines, and solar panels. But remember, by 2030, 30% of the land is to be preserved, and by 2050, 50% of all land is to be set aside. Manufacturing plants without gas-powered backup sources of electricity must run when the electricity is available, so more WWS systems must be built.

Without the barrel of oil lost because of the electric vehicles, there is no choice but to "electrify everything" that cannot be done with corn or another crop. At which point, there are only three choices for the end game:

1) The economy continues to grow, and the population increases, which means the electrical demand grows, so an ever-increasing demand for land for electrical generation, each successive plot of land being less and less windy or sunny, and land to grow crops for fuel, fibers, feedstocks, and feed while trying to put more land aside for preservation. Eventually, the country runs out of land. Land for food, fuel, fibers, medicine, and electrical generation, along with housing for the growing population. Famines, power outages, and overcrowding play out as the end results.

2) Land usage is treated as the limiting factor. The result is that the economy must be held back and population growth kept in check so that electrical demand will reach a steady state until technology comes along to find a way to provide the needed power. This scenario plays out with economic stagnation and a country that loses out in the marketplace to its international competitors.

3) Finally, America can choose to go it alone down the path of the green agenda without all the other nations following suit. The result is that America becomes increasingly dependent on other nations for all the fossil fuel derived products, giving up all its manufacturing and production. America quickly loses its technological and military superiority, facing economic stagnation and blown by the winds of world opinion.

Unfortunately, all the scenarios lead to dark futures.

The Problem with Computer Models

Political scientist Phillip Tetlock expressed his opinion on the validity of expert predictions by saying that the average expert's predictions were "roughly as accurate as a dart-throwing chimpanzee." I will contend that their computer models are equally accurate, or inaccurate as the case may be.

Have you ever noticed how behind every argument from climate change to pandemics is a computer model or simulation that the experts use to prove their theories? All predictions of the future must be based on something, and it is probably unrealistic for the experts to stand by their opinions as proof, so they use the computer model. It should be evident to everyone, anything run on a computer must have some science behind it and should be valid, right?

As Eric Dennis stated in *The Skeptic Smear*, "But what these [media] stories leave out is the evidential status of these developments-what any given study or model proves. And the answer is, little to nothing, because the present ability of scientists to understand, model and predict the climate is far, far lower than we are led to believe."[228]

Generally, in the engineering world, computer models are quite successful and valuable. Bridge designers use them all the time, and for the most part, the results are accurate and allow for the safe construction of bridges. We drive across bridges every day without questioning their safety, even if crossing major rivers or 20 stories in the air. Likewise, from car design to semiconductor performance, these engineering computer simulations use real-world results as feedback to continuously improve the model's performance. This is where any future computer model tends to run into problems. There is no feedback loop to make the model more accurate. It starts and ends with the coder's assumptions.

Computer simulations are only as good as the data provided. In the Florida International University pedestrian bridge collapse, the engineer who set up the computer simulations did so only in its final constructed state and not in the intervening partially constructed positions. As a result, on March 15, 2018, as the bridge was being put into place over a busy street and before the last connections could be made, the bridge failed, resulting in 6 deaths and 10 injuries. The NTSB report's outcome was simple: bad simulation, bad results.

The 2016 Presidential Election should provide a warning to everyone about overly trusting computer models. All the computer models had Hillary Clinton winning the presidency easily, in a landslide in some cases. Yet, all those models proved incorrect due to flawed assumptions about turnout and about party crossover. These models were highly funded projects by major media networks with real reputations on the line. How accurate can we then expect low-funded, obscure studies of which few will ever read and evaluate the results?

Anyone paying attention at the beginning of the Coronavirus pandemic in February to April of 2020 would be familiar with the plot on Figure 55. At nearly every press conference from the President to almost all the Governors, these two curves were shown to indicate that we had to lock down the country/world or the healthcare system would be overwhelmed. These curves were the product of a computer model written by Neil Ferguson of the Imperial College in London. Before the code was modified to be used for the Coronavirus, it had previously been used to predict that up to 150,000 people in the United Kingdom would die from mad cow disease (BSE) from beef, and in reality, there have been only 177 actual deaths.[229] When first released, the Imperial College model indicated 2.5 to 3 million Americans and ¾ of a million Brits would perish from Coronavirus. The model's fundamental flaw was that it failed to consider any of the healthcare community's ability to learn. The charts used a fatality rate of 0.95 for the entire plot. In April 2020, Stanford doctors Eran Bendavid and Jay Bjattacharya wrote in the *Wall Street Journal* that "current estimates about the coronavirus fatality rate may be too high by orders of magnitude."[230] A piece in *The Hill* reported the fatality rate was known to be "somewhere in the 0.1 to 0.2 percent range"[231] as early as April 22, 2020. In reality, the average value over 2020 was 0.26, but it ranged from 0.9 in the spring to less than 0.05 by the end of the year. By early May 2020, just two months after the modeled curves were revealed to the world, Dr. Ferguson resigned in disgrace due to the sloppiness of his projections and his own unwillingness to abide by the lockdown rules. Yet, while the actual numbers were understood before Dr. Ferguson's resignation, those curves are still being used to justify lockdowns more than a year later. That same model has been used to predict the effect of climate change on the population.

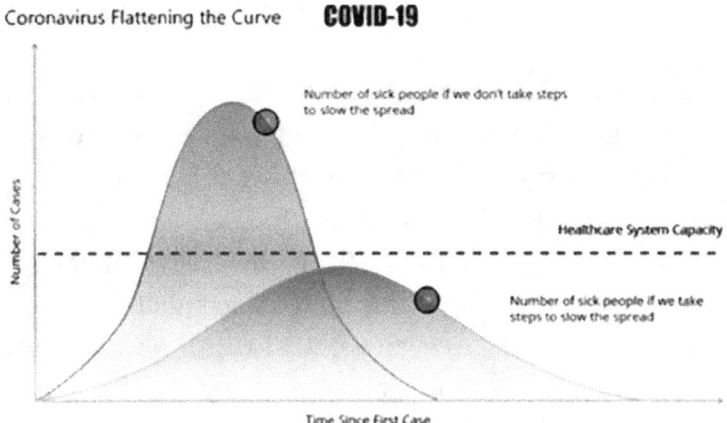

Figure 55 Imperial College Covid Computer Model (Spring 2020)

Even models predicting near-term events with understood disease mechanisms have failed to be accurate. Every year, we face influenza, which spreads in the same manner and has similar traits.

Yet, it seems models fail to accurately predict behavior over the next few weeks, let alone decades into the future.

Similarly, the computer models for climate change lack the necessary feedback loops and the required information about the mechanism to ensure that they are correct. Eric Dennis points out that without this data, "there is a lot of room here for the ultimate outcome of the models to be controlled by ideological predispositions" since these simulations are "model code that they control and have played with for years."[232] Traditionally, science happened in isolated labs, mainly utilizing academic funding, with the intended outcome being academic paper generation. Increasingly, science has become expensive. The electron was discovered in 1897 with the use of a few hundred dollars' worth of equipment. But, in 2020, the European Organization for Nuclear Research (CERN) asked for $15 billion to build a collider to discover the next elementary particle.[233] The only manner in which such large amounts of funds can be raised is through corporate sponsors. As such, what once was the realm of academia is now the realm of corporate marketing.

In 1968, 100 current and former heads of state, UN officials, scientists, and business leaders got together in Rome to create an organization to suggest solutions to man's problematic issues: the environment, poverty, crime, and health. Their key work, **LIMIT TO GROWTH** which was published in 1972, is the basis of so many of the predictions of the 1970s, along with its computer models and projections. Fifty years on, even with being fully aware of the lack of sophistication of those models and seeing the real-life results that failed to be verified, surprisingly, those predictions continue to be used by some activists today.

As a real-life example of where modeling can go wrong, the ABC network filmed a program in 2008 called "Earth 2100," which was broadcast in 2009.[234] They declared that by 2016 milk would be

$12.99 a gallon and gasoline over $9 per gallon. A further clip showed New York City engulfed in water, as their model predicted a substantial rise in the sea level. If they were that wrong over just 7 years, how accurate can they be over 100?

Dr. William Happer, the Cyrus Fogg Brackett Professor of Physics, Emeritus at Princeton University, gave a talk in February 2021 where he discussed some of the most recent climate models.[235] He noted that the actual measured readings, while increasing, have not followed the models upon which so much of our planning is based. In a study, the results from many runs of the IPCC simulations, a total of 102 of them, were averaged and plotted. In Figure 56, this curve of increasing future temperatures is compared to actual readings of the earth's climate. While the recorded measurements over the past 20 years are ever-growing, it is never by the amount of the runaway values in the IPCC curve.

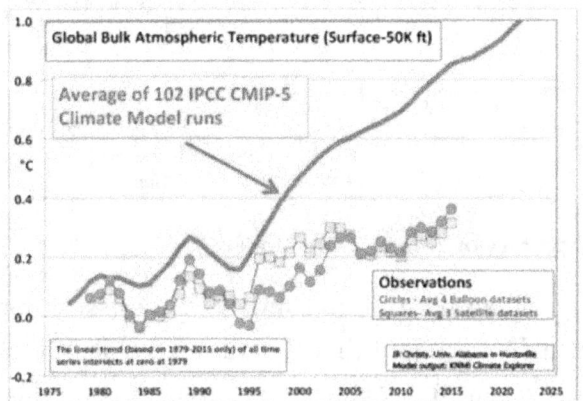

https://science.house.gov/sites/republicans.science.house.gov/files/documents/HHRG-114-SY-WState-JChristy-20160202.pdf

Figure 56 Runaway Temperature of the IPCC Predictions

He further discussed the diagram in Figure 57, which shows the results of hundreds of simulations on computer models to determine the expected temperature increases. The single tall white column represents the actual measured results in the same time frame. The computer models repeatedly overestimated the anticipated temperature rise. This diagram comes from *Nature – Climate Change*, the recognized authoritative journal on climate science. These computer models struggle with estimating the impact of clouds on temperature and the self-correcting nature of clouds as more heat means more clouds, while the presence of clouds can have a cooling effect.

Modelled warming (gray) is much larger than observed warming (red).
a
b
Happer et al., 1982
Normalized density
0.0
0.2
0.4
0.6
Change in temperature
(°C per decade)
Figure 1 | Trends in global mean surface temperature. a, 1993-2012. b, 1998-2012. Histograms of observed trends (red hatching) are from 100 reconstructions of the HadCRUT4 dataset'. Histograms of model trends (grey bars) are based on 117 simulations of the models, and black curves are smoothed versions of the model trends. The ranges of observed trends reflect observational uncertainty, whereas the ranges of model trends reflect forcing uncertainty, as well as differences in individual model responses to external forcings and uncertainty arising from internal climate variability.
Fyfe et al., Nature Climate Change, Vol 3, p. 767, September 2013.

Figure 57 Inaccuracy of Temperature Models

One of the problems with climate models is the assumption that CO_2 will be unusually high and that this high level is not conducive to life. A premise which actual measured history shows to be otherwise. The fact is that the current CO_2 is at a near-historic low when discussed over a timeframe of hundreds of millions of years. The 20-year-old plot in Figure 58, originally published in 2001, shows the historical level of CO_2. While recent CO_2 levels are higher than other times in the past century, it was even higher during most of the earth's history. The computer models seem to miss the fact that life did not die off. The minimum CO_2 level which will sustain

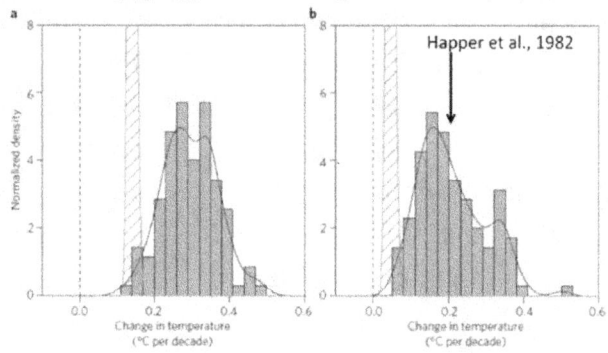

plant life is roughly 150 parts per million. While we currently sit around 415 ppm, the environmentalists would like to get the world under 350 ppm. As Gosselin writes on his website NoTricksZone, "Worrying that 400 ppm is too high is like worrying about your fuel tank overflowing when it reaches the 1/8 mark during filling."[236] Science has shown that plants begin to suffer when CO_2 falls below 500 ppm, which is why greenhouses use CO_2 emitters to improve plant growth.

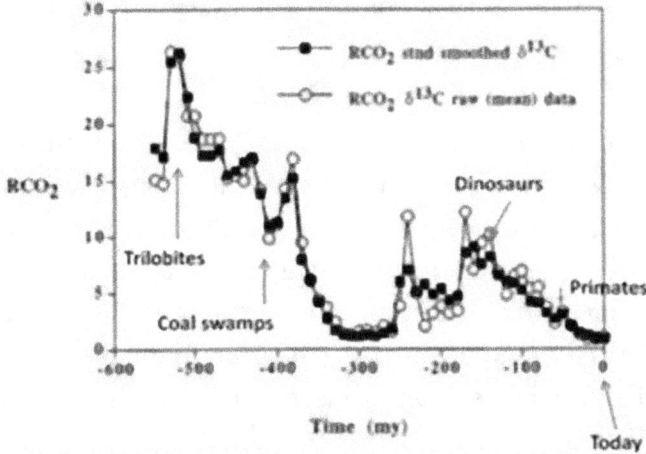

Figure 58 Historic CO_2 Levels

Care must be taken with any consideration of computer models that claim extreme changes over limited time periods. Past simulations and models have proven highly inaccurate. Using actual measurements to compare to early models has been disheartening. Still, we are promised that the new models are more sophisticated and will be much more accurate. But is there evidence of this? Beware of magical thinking when looking at models, are the results based on science or the author's ideology?

10

The Dark Future

> *"If we don't address those two issues – of climate change and growing inequalities – we will be moving towards a dark future 50 years from now."*[237]
>
> Christine Lagarde, Head of the International Monetary Fund, 2017

At a 2017 conference, Christine Lagarde, the head of the International Monetary Fund (IMF), commented that if actions were not taken soon, the world would be "toasted, roasted and grilled,"[238] and would face a very dark future. If we wanted a world that would "look like utopia and not dystopia," then the world would need to stand on the right side of history and tackle the issue of climate change. Climate change, that is, as framed by activists and the United Nations.

In a short video produced in 2020 by the World Wildlife Fund (WWF), Sir David Attenborough laid out how he viewed the plan for saving the world. He had four commandments: "1) stop the damaging stuff, 2) new green technology, 3) get the population down and 4) keep hold of the natural wealth we currently got."[239] Basically, his plan boils down to using new technology to reduce the number of "undesirables" and keep the nations' relative positions in place. Those that "have" get to keep it, and those who "don't have," don't get it. Sir Attenborough finishes his video with the line, "We now have the choice to create a planet that we can all be proud of." Is the planet of which Sir Attenborough would be proud be one which we would want to live on?

Activists like Sir Attenborough define the current geological epoch by the term Anthropocene, meaning *"The present geological time interval, in which many conditions and processes on Earth are profoundly altered by human impact. This impact has intensified significantly since the onset of industrialization, taking us out of the Earth System state typical of the Holocene Epoch that post-dates the last glaciation."*[240] The use of this term is intended to indicate that the effects of man are greater than the natural powers of the planet. The starting point for this period is debated but placed somewhere between the beginning of the industrial revolution and the 1945 introduction of atomic power.

The Final Ten Years

It is always ten more years. It seems that every climate activist has said, "we have 10 (sometimes 9, sometimes 12) more years to change the world, or the world will come to an end." It does not matter if it is a President (Barak Obama, Joe Biden), government officials (Bernie Sanders, Alexandria Ocasio-Cortez, John Kerry), economic activists (Tom Steyer, Bill Gates), environmental activists (Greta Thunberg, Bill Nye), religious leaders (Pope Francis) or international bodies (IPCC)[241]. They all say the same thing, time is short, or the end is nigh.

In an April 12, 2021 article, *Scientific American*[242] announced we were no longer threatened by "climate change" but that it had become a "climate emergency," a new moniker that everyone should be using. The article claims that 13,000 scientists have agreed that this is the proper wording claiming, "we are on solid scientific ground" and that this was "not a journalistic fancy." This follows the Oxford dictionary declaring "climate emergency" as the word of the year in 2019, although I would hate to point out to such a prestigious source that it is a phrase, not a word. People magazine[243] even carried the announcement noting that "Journalism should reflect what science says: the climate emergency is here," concluding that more than 200 publications and news outlets signed on to follow suit in the word's usage. But this is not the first climate emergency.

I remember the first time I encountered the "ten years is all we have left" scenario. The pushing of the environmental gloom and doom within the classroom is not a new phenomenon but has been going on for nearly all of my lifetime, more than 50 years.

I turned 9 years old, the spring of the first Earth Day in 1970. I remember, and the old school papers still exist to prove it, that we spent most of the spring semester of fourth grade studying pollution and the ill effects it would have on the world. Major publications and magazines were trumpeting the coming ice age and the fact that we would all freeze to death soon. In the January 1970 issue of Life Magazine, a story ran indicating that with greenhouse gases being as they were, "...by 1985 air pollution will have reduced the amount of sunlight reaching earth by one half"[244]. The result of this reduction of sunlight would be, of course, cooler weather.

On page 64 of the April 28, 1975 issue of Newsweek magazine, an article entitled "The Cooling World" noted that "There is ominous signs that the earth's weather patterns have begun to change dramatically and that these changes may portend a drastic decline in food production" and that "the drop in food output could begin quite soon, perhaps only ten years from now." There it is again, those ten years till the end. The cause of the problem was seen as "a drop of half a degree in average ground temperatures in the Northern Hemisphere between 1945 and 1968" and "a sudden, large increase in Northern Hemisphere snow cover in the winter of 1971-72." If there was one thing clear from the scientific facts and the words of the scientists, that by 1985, the world would be irretrievably freezing and that "the longer the planners delay, the more difficult will they find it to cope with climatic change once the results become grim reality." It has been 45 years, and we are still anticipating that grim reality. What is the cause of all of these changes and the cooling earth? Greenhouse gases and aerosols, of course, which were going to block the light of the sun.

The writer of the Newsweek article, frustrated at its repeated use in this very context as a "see they got it wrong" report, wrote a follow-up piece for *Inside Science* entitled, "My 1975 Cooling World Story doesn't make Today's Climate Scientists wrong." In the article, he writes that the error was that today we use "theoretical concepts that fit into computer models and an overall framework outlining the nature of Earth's climate. These capabilities were primitive or non-existent in 1975." He believes that the difference between the old global cooling and current global warming is that our air is cleaner since removing sulfate aerosols, which had a cooling effect. Interestingly, his defense of the old article is that we did not know enough, and nothing says, "climate science never changes." Today, he is certain that climate science has it right this time because of a new climate field called 'detection and attribution,' whose sole purpose is to find problems and attribute such effects to human activity. He concludes that using his article as evidence is part of the "American discourse that is anti-intellectual and anti-expertise."[245]

In fact, the world has been serially doomed for many years. UK scientist Philip Stott points out the phenomenon when he explained, "In essence, the Earth has been given a 10-year survival warning regularly for the last fifty or so years. We have been serially doomed." If anything, he has been too generous as it has been going on for far more than 50 years.[246]

George Perkins Marsh may have been the first "gloom and doomer" when in his 1864 book, **MAN AND NATURE; OR, PHYSICAL GEOGRAPHY AS MODIFIED BY HUMAN ACTION**, he said, "The earth is fast becoming an unfit home for its noblest inhabitant, and another era of equal human crime and human improvidence ... would reduce it to such a condition of impoverished productiveness, of shattered surface, of climatic excess, as to threaten the depravation, barbarism, and perhaps even extinction of the species."[247]

The United Nations has been trumpeting the gloom and doom agenda for some time, having hosted their first World Climate Conference in Geneva in 1979. In the first Intergovernmental Panel on Climate Change (IPCC) release in 1990, the UN reported, "entire nations could be wiped off the face of the Earth by rising sea levels if the global warming trend is not reversed by the year 2000."[248] The report went on to give the standard "10-year window of opportunity to solve the greenhouse effect before it goes beyond human control." Noel Brown, director of the New York office of the UN Environment Program or UNEP, said that the most conservative scientist "already tells us there's nothing we can do now to stop a change of" between 3 and 7 degrees within the next 30 years.

Within the last decade, the predictions of an ice-free arctic have been popular. The ice-free arctic, which never happened on time, has been predicted by Professor Wieslaw Maslowsi, (melted by 2013), NASA's Jay Zwally (2012), Professor David Barber (2008), NSIDC Director Mark Sereezer (2030), Professor Peter Wadhams (2015), and "hundreds-more dire Sea Ice Predictions."[249]

As the predictions of gloom and doom continue to be published, it might be worthwhile to review some of the past predictions. Publications have been using the threat of climate change to sell papers for more than a century. Here is a quick list of predictions

The Dark Future

that were published long before most believed that this was an issue.

1895	Geologist think the world may be frozen up again – New York Times
1902	Disappearing Glaciers – Los Angeles Times
1912	Prof. Schmidt Warns Us of an Encroaching Ice Age – New York Times
1923	Scientist says Arctic ice will wipe out Canada – Chicago Tribune
1924	MacMillan Reports Signs of New Ice Age – New York Times
1929	Most geologists think the world is growing warmer and that it will continue to get warmer – Los Angeles Times
1932	If these things are true, it is evident, therefore, that we must be just teetering on an ice age – The Atlantic
1933	America in Longest Warm Spell since 1776, Temperature records a 25-year rise – New York Times
1938	Global Warming, caused by man heating the planet with carbon dioxide – Royal Meteorological Society
1939	Weathermen have no doubt that the world, at least for the time being, is growing warmer – Washington Post
1952	We have learned that the world has been getting warmer in the last half-century – New York Times

There were four changes of opinion in the 90 years between 1900 and 1990, twice to the earth is cooling and twice to its warming.

With thanks to Myron Ebell and Steven J Milloy of the Competitive Enterprise Institute, and numerous others who have expanded on the list, here is a list of more than 50 gloom and doom predictions that failed to materialize.[250] It is interesting to note that the projections are clustered from 1966-1980 (primarily global cooling) and 1996-current (global warming). Somehow, from the years 1980 to 1996, there was significantly less gloom and doom of climate change, precisely those years in which the United Nations began their efforts.

Table 13 Historical Record of Failed Predictions

Year	Prediction
1966	Oil Gone in Ten Years
1967	Dire Famine Forecasted By 1975
1968	Overpopulation Will Spread Worldwide
1969	Everyone Will Disappear In a Cloud Of Blue Steam By 1989
	Worldwide Plague, Overwhelming Pollution, Ecological Catastrophe, Virtual Collapse of UK by the end of 20th Century
1970	Ice Age By the year 2000
	World Will Use Up All its Natural Resources
	America Subject to Water Rationing By 1974 and Food Rationing By 1980
	Urban Citizens Will Require Gas Masks by 1985
	Nitrogen buildup Will Make All Land Unusable
	Decaying Pollution Will Kill all the Fish
	Killer Bees!
	Oceans Dead in a Decade
1971	New Ice Age Coming By 2020 or 2030
1972	New Ice Age By 2070
	Oil Depleted in 20 Years
1974	Space Satellites Show New Ice Age Coming Fast
	Another Ice Age?
	Ozone Depletion a 'Great Peril to Life
	Pending Depletion and Shortages of Gold, Tin, Oil, Natural Gas, Copper, Aluminum
1975	The Cooling World and a Drastic Decline in Food Production (the Newsweek Article noted above)
1976	Scientific Consensus Planet Cooling
	Famines imminent
1977	Department of Energy Says Oil will Peak in the 1990s
1978	No End in Sight to 30-Year Cooling Trend
1980	Acid Rain Kills Life In Lakes
	Peak Oil In 2000

1988	Regional Droughts in the 1990s
	Temperatures in DC Will Hit Record Highs
	Maldives Islands will Be Underwater by 2018
	World's Leading Climate Expert Predicts Lower Manhattan Underwater by 2018
1989	Rising Sea Levels will Obliterate Nations if Nothing Done by 2000
	New York City's West Side Highway Underwater by 2019
	UN Warning: Entire Nations Wiped off the face of the earth by 2000 from Global Warming
1996	Peak Oil in 2020 (well, this may prove to be true, eventually)
2000	Children Won't Know What Snow Is
	Snowfalls are now a thing of the Past
2002	Peak Oil in 2010
2004	Britain will Be Siberia by 2024
2005	Manhattan Underwater by 2015
	Fifty Million Climate Refugees by the Year 2020
2006	Super Hurricanes!
2008	The Arctic will Be Ice Free by 2018
	Climate Genius Al Gore Predicts Ice-Free Arctic by 2013
	New York City Underwater by 2015
	Miami Ocean Front Flooded by 2015
2009	UK Prime Minister Says 50 Days to 'Save The Planet From Catastrophe'
	Climate Genius Al Gore Moves 2013 Prediction of Ice-Free Arctic to 2014
2011	Cherry Blossoms Blooming in Winter
2013	Arctic Ice-Free by 2015
2014	Only 500 Days Before 'Climate Chaos'
2018	The last Winter Olympics due to lack of snow

The activists are consistent: doom is always right around the corner and even if it fails to come true, keep pushing forward on the warnings. A rational person would see that the doom did not happen and reevaluate the methods to understand what was missed and how to fix the forecast. However, as this never seems to occur to those making the predictions, maybe you should question the purposes of the methods.

The pushers of the gloom and doom forecasts seem to fall into two main categories, those who intend to make a career of the doom and those who intend to make a legacy of the collapse. Climate change may fall into that category, like so much of politics, that if you are not for it when you are young, "you don't have a heart," and if you aren't opposed to it when you are older, "you don't have a brain." At least, that is what the statistics among age groups go to show. Although all age groups claim the same amount of knowledge (which is undoubtedly overestimated), the older the group, the less concerned they are about the issue. Do not forget that these "older" groups were the younger groups during the previous periods of climate change crises.

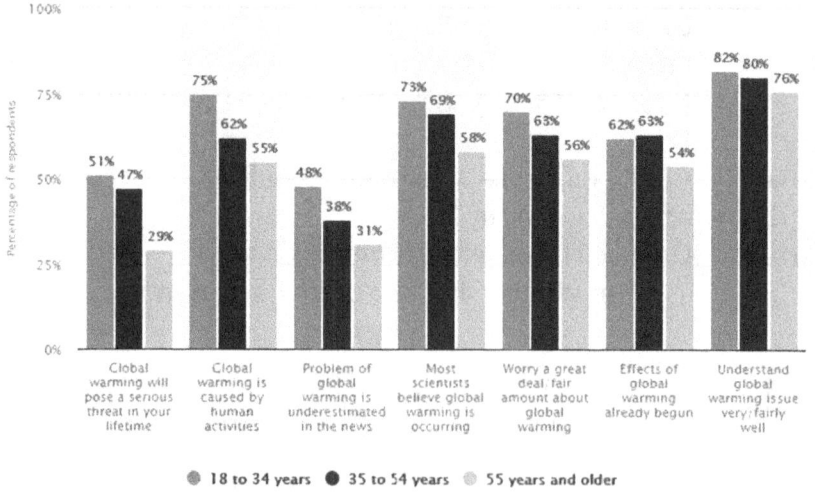

Figure 59 Concerns about Climate Change[251]

It is important to remember that so many gloom and doom industry stories are academic research primarily funded because they find something to report. Any study that would conclude, "we are good, don't worry about it," will never get a second round of funding. These academics require two things to maintain their academic career: publication and funding.

Either way, when evaluating the validity of any argument made by these individuals, one must consider whether their personal actions align with the predictions for which everyone else's lives should change. During his time in office, President Obama took many steps in the name of climate change, from resisting the Keystone XL pipeline to signing onto the Paris Climate Accord in 2015. Upon retirement from the White House, he soon purchased a $15 million oceanfront mansion on Martha's Vineyard.[252] The research group, Climate Central, using the NOAA's database, concluded that Obama's mansion was listed as "unlikely, but increasingly plausible" for flooding due to climate change. When Obama puts his money on the question of whether the oceans are rising due to climate change or not, he clearly bets on them not rising.

Likewise, former Vice-President and later environmentalist Al Gore used the money from his movie, "An Inconvenient Truth," to purchase a $9 million gated ocean view villa on 1 ½ acre in Montecito in Southern California.[253] Again, his dollars voted against the oceans rising.

Finally, President Biden's special envoy for climate, John Kerry, reported on his financial disclosure form that he invested $4.2 - $15 million in oil and gas companies in the month before taking the climate czar role. Furthermore, Kerry has his $11.75 million historic property at Chilmark on Martha's Vineyard[254]. A property for which in 2009 he, along with Representative Ed Markey (D-Mass.), and Senator Ted Kennedy and the Kennedy family, spent significant amounts of money to fight the Cape Wind (later renamed Vineyard

Wind) 130-turbine wind farm to be located 8 miles offshore.[255] Kerry's belief in climate change is so strong that he refuses to have wind turbines disturb his view while owning a home that would be flooded if the oceans were actually to rise. Clearly, his personal beliefs are showing.

Maybe President Biden is the most respectful of the rising oceans as his Rehoboth Beach House, which he purchased in 2017 for $2.7 million, is a "couple of blocks" from the oceanfront. With the nominal ocean rise, the beach house becomes oceanfront property, but an above nominal rise in the sea could flood the home. Still, his primary home with 6850 square feet of living space for two people, along with a cottage to rent to secret service personnel, does result in a significant carbon footprint for the President.

Each February, world leaders meet in Davos, Switzerland, to confer about the day's various crises. Though they are rich and powerful, having made their money through fossil fuels' benefits, they overwhelmingly support the climate change theory. They often spend their days discussing how the earth's mean temperature can be lowered. One might find it surprising to know that these climate proponents used more than 1,500 individual private jets to attend the conference in 2019,[256] and as the *Wall Street Journal* reported, "to reduce the event's carbon footprint, no paper maps of the town were being distributed; one could almost feel the waves of relief from the nearby Alpine glaciers at this sign of green progress."[257]

But it isn't just the rich and famous that struggle with hypocrisy. During the 2018 Winter Olympics and afterward, American gold medalist cross-country skier, Jessie Diggins, lamented that this would be the last winter Olympics as in just a couple of years, there would be no more snow, or at least not enough to hold a Winter Olympics. What was Jessie's response to the melting snow? Not much actually, she has continued to train and in 2020 won the FIS cross-country world cup title and two additional medals in the 2022

Winter Olympics. Dedicating your life in pursuit of an activity that you believe would not exist in a couple of years would be foolhardy. Yet, she continues to warn of the disappearing snow in between her hours on the snow-covered hillsides.

Here locally on Saturday morning radio, the local car guy touts the ill-effects of climate change and the need to abandon fossil fuels, all while spending more than 30 years making money first by selling and later promoting gas-burning automobiles. His business model hasn't changed despite the increased concern over the environment. Sometimes, the obvious isn't so obvious to those involved.

What is clear is that when the current crop of doom and gloomer's move on, there will be more to follow suit. As an article from *Electroverse* points out, "I wonder what new power-hungry know-nothing, know-it-all politician we'll be hearing from in 2031, when AOC's prediction also uneventfully passes us by."[258] But then what enjoyment is there to be virtuous and politically correct if no one knows about it?

Are Lockdowns here to stay?

In response to the coronavirus pandemic panic, the developed nations decided to shut down their societies, putting their economic futures at risk. These lockdowns were a financial luxury, afforded only by the wealthiest nations and brought into a possibility by the advanced technologies that fossil fuels had enabled. The top half of the socio-economic stratum drove the demand for a lockdown as their ability to "work from home" was powered by non-labor intensive jobs. All those jobs, freed from the arduous tasks of agriculture and manufacturing, created by the economic boom years that only required computer access to accomplish. All that high-tech gear built and powered with fossil fuels made zoom calls and virtual schools possible. Imagine a world of lockdowns before high-speed internet and video teleconferencing.

On Feb 26, 2021, the World Economic Forum posted a video entitled, "Lockdowns are quietly improving cities around the world." They quickly realized that the message was going to be taken wrong and recut and reissued the video. The video exclaims such extraordinary achievements as "Earth's quietist period in decades" and that "Urban ambient noise fell by up to 50%... as buses and train services were reduced and aircraft grounded, and factories shuttered."[259] In the revised cut, they trumpeted the "value to seismologists of the quiet for earthquake detection."

Of the many unintended consequences of the covid lockdowns of 2020, many will have long-term effects, especially when considering the possible retirement of fossil fuels. Society changed in many ways during the lockdown from travel patterns, work locations, and medical care access. According to the World Economic Forum video on the lockdown's benefits, the world's greenhouse gas emissions dropped 7% during the year. This could be considered a positive move, although the yearly goal is a 7.7% reduction for each of the next 10 years. An article in *Nature – Climate Change* suggests that the world should consider economy-wide lockdowns of this type every other year to maintain the temperature limit goals.[260]

Many would review the results of the covid lockdowns as unintended consequences. There are essentially three groups of people when it comes to thinking about the unintended consequences of decisions. First, the average person makes decisions, and even when they research and do the best they can, there can be unintended consequences because they are rarely experts and have limited time and resources. The second tier of decision-makers is represented by most members of our federal and state legislatures. They have access to more research and information but again are not likely to be experts in the field and have limited time to apply to a single bill or proposal. While there may be some unforeseen results from their actions, they should be considerably fewer. However, the third group is those whose jobs it is to know these things. Their job assignments are to know the facts, have access to all the information, and are paid to consider all the alternatives. In this context, these people work for lobbying firms, energy companies, and vendors selling products. With this group, there are no unintended consequences. There is nothing they haven't considered, although whether they act on that knowledge can be a different matter.

The difference between a conspiracy theory and a fact is whether it can be proven. Some look at the World Economic Forum's published material to point out that the consequences were not unintended at all. That may be the plan was to disrupt how society lives. Whether the lockdowns were enforced so that we would move to new patterns is an open question, but the fact remains that indeed new life patterns have emerged from the lockdowns.

Workers have learned to work from home, and increasingly the office arrangements look more like shared office spaces with fewer people having their personal reserved desks. Most residences are heated and cooled, irrespective of the presence of the residents. Thus, the shared desk concept is overall a green policy as it utilizes all that residential energy that otherwise would go to waste.

The 2020 lockdown resulted in a massive reduction in travels of all forms, car journeys down 26%, mass transit down 63%, and ferry rides down by 60%. Air travel demand fell by 66% in North American and 70% in Europe.[261] And as a result, airlines worldwide saw their profits decline by $118.5 Billion. In 2019, 4.5 billion passengers took to the air, which fell to 1.8 Billion during the pandemic. All those flights that were not needed meant that all that aviation fuel was also not required, a good thing when the gasoline wasn't burned on car journeys or diesel burned by the mass transit buses.

Interestingly, while bicycle sales were up 81%, mostly for virtual schooling children, the rental of bike-share and scooters was down so significantly that 45% of bike-share and 64% of scooter operations went out of business during the year.

As the higher portions of socio-economic society pushed for the lockdowns, the other half of society suffered. While all those freed from manual labor could work from home, those not so lucky could not. Without all those workers going to the office, the demand for clothing declined. In fact, Bangladesh who saw synthetic fibers drive its economy saw an 85% reduction in clothing exports in April 2020. The manual laborers in developed nations send more money home, $554 billion, "than all foreign direct investment in low—and middle—income countries and more than three times the development aid from foreign governments."[262] With unemployment rates reaching 40%, those workers who perform manual labor sent 23% fewer dollars home to countries like India, the Philippines, Mexico, and El Salvador.[263] The lockdowns of the wealthy nations brought increased poverty to the lesser developed ones. Arif Husain, the chief economist of the United Nations' World Food Programme, projected in April 2020 that 285 million people could be pushed to the brink of starvation by the end of 2020.[264] If lockdowns become a regular event, then starvation will become a recurring one.

To accomplish the reductions of greenhouse gas emissions for which the **Green Solution** calls, it will require adjusting to life as if the world never emerged from the covid lockdowns. Now that we have adapted to life in lockdown mode, we have made the first step towards living in a low-energy culture. This stage is to live life with intermittent energy, without the devices upon which so many depend, and without all those life-saving technologies. With social lockdowns, many people are experiencing their first insights into the dark future of a fuel-free world.

The Mission to Cut Emissions

On Earth Day 2021, President Biden held a virtual climate summit at which he pledged to cut emissions in half by 2030. It was reported that "The White House's goal of reducing greenhouse gas emissions by 50% to 52%, from a baseline of 2005 emissions, is nearly double the target set by the Obama administration in 2015. An administration official, who briefed reporters on the condition of anonymity, did not detail how the White House plans to achieve the 50% reduction in emissions."[265]

The renewable grid ideas discussed in Chapter 9 were developed by teams of experts, who presented plans by which the nation could move to these mostly renewable grids by 2035 or 2050. So why did President Biden's goal go to 2030? Well, it's those ten final years again. Once again, we have ten years to save the planet or doom. The experts needed the extra years for several reasons: the technologies need time to develop as they aren't ready today, only so much hardware can be built in a year, and there is only so much money to spend. However, as Thomas Sowell so elegantly put it, "a rudimentary knowledge of economics is not a requirement for a career in politics [or] journalism ... Certainly, it is not a prerequisite for colorful expressions of moral indignation."[266]

It might seem strange to the casual reader that the President has a goal for 10 years in the future but places the starting line 15 years in the past. There is a good reason for it, and that goes back to the 1997 Kyoto Protocol that went into effect when it was ratified in 2005. The treaty provides emission reduction targets for 37 industrialized countries and the European Community. It is also a beneficial starting place as it represents the peak date for U.S.

emissions at 7,400 million metric tons of CO_2 equivalent, of which 82% were CO_2.

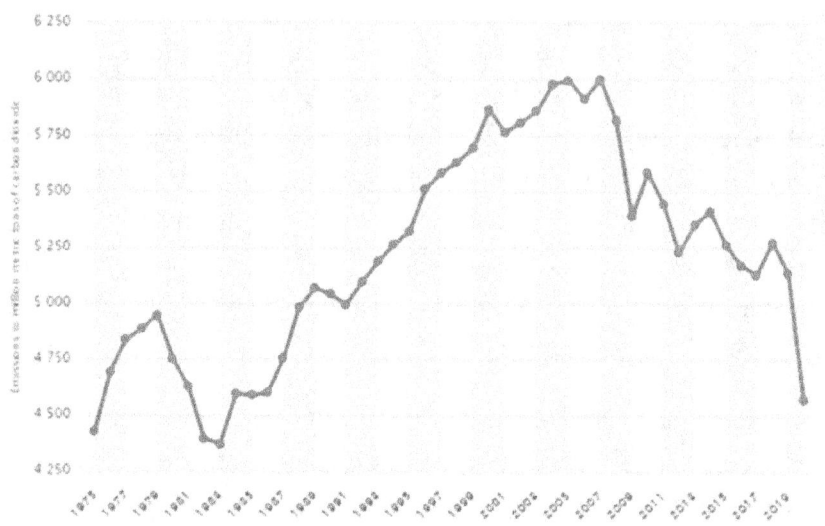

Figure 60 CO_2 emission from Energy Consumption, U.S. 1975 - 2020

Luckily for President Biden, he gets a head start to his goal of 50% reduction as he takes credit for improvements that have already happened. With the help of the 2008-2009 recession, emissions have fallen by an average of 1% per year, resulting in a 13.6% head start when comparing 2005 with 2019. However, the 2020 lockdown added another 10.3% to the drop, which means that the 2021 starting point to Biden's promise is down 21% below the 2005 levels. However, 2021 emissions will likely be significantly higher than the level for the previous year unless lockdowns continue.

Table 14 2019 U.S. Emissions Breakdown

Economic Sector	Percent of Emissions	Percent of Category	Change 2005 to 2019
Transportation	28.6%		-5.1%
Light-Duty Vehicles		59%	

Medium and Heavy Duty Trucks		23%	
Aircraft		9%	
Other (Bus, Motorcycle)		5%	
Rail		2%	
Ships and Boats		2%	
Electricity Generation	25.1%		-32.9%
Coal		62%	
Natural Gas		37%	
Petroleum		1%	
Industrial	22.9%		-0.9%
Agriculture	10.2%		+6.3%
Commercial	6.9%		+11.6%
Residential	5.8%		+2.3%

The White House provided no ideas about getting the nation to that magical 50% point, nor what the consequences to the citizens would be. In the April 23, 2021 online version of the *Wall Street Journal*, a clever web app was presented that allowed the reader to select from several choices upon which to decide the final 36.4% of cuts in emissions. Choices included increasing renewable energy, switching to electric vehicles, replacing diesel buses, electrifying buildings, eliminating fracking, and closing coal-fired plants. The game was rigged though, as it had been predetermined that you had to select a carbon tax to get there and either use renewables or all the other options. So the fix was in, and the article appears to merely be a means to convince the readers of that.

The Center for Global Sustainability at the University of Maryland created a project to determine how to reduce by 51%, and its result requires electricity emission cut by 76%, transportation by 40%, industry by 16%, and buildings by 18%. Their proposals are:

- 50% renewable electricity

- 65% of light-duty vehicle sales and 10% of heavy-duty truck sales be electric vehicles
- All new buildings 100% electric
- Cut natural gas leaks by 60%
- 20% more forest

Of the actions that could be taken to reduce emissions by 2030, three areas are too slow to respond in time: buildings, carbon sinks, and industry. The median age of a home in America is 37 years ranging from 58 years in the northeast to 25 in the southwest. While new ideas can be incorporated in new builds, there simply aren't enough new builds to make a significant difference in just 10 years. The same is true for commercial buildings that have a typical lifespan of 35 years. The second area, carbon sinks, involves the growth of denser trees and other plants designed to hold the carbon for decades to come. These plants grow just too slow. And finally, industry takes time to develop, implement and commercialize new low emission techniques, too much time to make a difference in just 10 years. So, the actions that can make the 50% reduction are limited to electricity and transportation, although even for these, the 10-year window is aggressive.

Achieving the reductions by 2030 means utilizing what we have available today. Within the realm of electricity generation, shutting down the coal plants is the obvious first step, replacing them with renewables, moving renewables from 20% to 40% of total generation capacity. Unfortunately, this replaces baseload, independent, controllable, and reliable sources with peak usage, interdependent, uncontrollable, and unreliable sources. So, building additional capacity will be necessary, and the public should develop an expectation of intermittent electricity.

The implication of moving to electric vehicles is discussed elsewhere but moving to the proposed date of 2030 largely requires the automobile manufacturers to significantly push the design and introductions of these models. As the CEO of Toyota, Akio Toyoda, said recently, if the world is too hasty in banning gasoline-powered cars, "the current business model of the car industry is going to collapse."[267] He further points out that "when politicians are out there saying, 'Let's get rid of all cars using gasoline,' do they understand this?" He was quoted as saying that "Japan would run out of electricity in the summer if all the cars were running on electric power." His statement was not even considering if the grid was relying upon renewables. In the end, the push to electrics will result in cars becoming a "flower on a high mountain," out of reach for the average person. But then maybe that is the idea.

The Great Dislocation

The great civilizations of ancient history, Rome, Greece, Carthage, Egypt, Persian, and India, all have one thing in common, they existed in a narrow band of moderate weather around the globe. Their seasons were not too hot nor too cold. While the Roman and Greek Empires bloomed, the northern parts of Europe were busy cutting down trees in an effort to keep warm in the winter. In the millennium between 900 and 1900, Germany's forests decreased from covering 70 percent of the land to just 25 percent.[268] In the past century, almost half of that loss has been recovered, thanks to not burning wood for fuel.

The introduction of intermittent energy by way of wind and solar renewables will have the consequence of encouraging individuals and corporations to escape those places where that energy would be dangerous. Northern winters are notorious for low sunlight and light winds, which is a disaster when the power is most needed for warming to avoid the extreme cold they experience. Therefore, expect the population's mass movement to the south to states like Texas and Florida, where intermittent power would not be life-threatening.

Due to regional political differences, there is already a significant population movement from the northern states towards Texas and Florida. The WEF famously predicts 50 million climate refugees if no action is taken on climate change. Still, possibly more refugees will exist if the energy is not available to keep warm in the winter and cool in the summer.

Europe's population will be caught in contrary issues. For years, migrants have been moving away from the poverty and conflict of the Mideast towards Italy, Germany, and Sweden. However, the intermittency of renewables may encourage those in Germany and Sweden to head south.

The Coming Health Crisis

In their paper, IPCC, "Making Peace with Nature," indicates their view of the future of medication. They state that "an estimated 4 billion people rely primarily on natural medicines for their health care, with communities living in lower-income settings particularly reliant on largely plant-based traditional medicines. The health of these people is compromised as wild-collected medicinal plants become less available. Some 70 percent of drugs used for cancer are natural or are synthetic products inspired by nature, and more than 20 percent of modern drugs used for all diseases are based on leads from natural molecules, identified by science or based on indigenous local knowledge, including aspirin, vincristine, and taxol. Though novel natural medicines are continuously being identified, the potential for future discoveries is critically undermined by biodiversity loss."[269]

While the IPCC focuses on the loss of biodiversity issues as a threat to future medications, they seem to miss the obvious problem with their proposed solutions, threatening the production of today's medications. The pharmaceuticals for America and the rest of the world are created from petrochemical feedstocks, of which 80%

originate in China. Without fossil fuels, those feedstocks are much more difficult to generate. Shut down the petrochemical industry, and there is a significant time lag before new medications based on new feedstocks can be readied and approved. On average, it takes 10 years and $2.6 billion to develop and release each new successful drug. The world without those chemicals will be significantly less healthy for at least those final 10 years.

Transportation

All the plans of shutting down the fossil fuel industry will leave the world with a dark future regarding moving people and materials worldwide. The world's economy is based on trade, and trade is based on transportation.

With its tighter emission standards and increased incentives, California has become the home of almost half of all the plug-in electric cars sold since 2011, a total of over 850,000. New York, Florida, and Washington are the second through fourth-ranked states, respectively, each with around 4% of total sales of EVs. While electric vehicles have been on sale for more than a decade, their acceptance still lags, and the consumer's experience is less than expected for the vehicle of the future. The two most commonly purchased electric vehicles have been the Toyota Prius and Tesla Model S, both models have risen to higher sales numbers as they became trendy, with the Prius sales disappearing as the Tesla took off. Basically, consumers are buying to be part of the fad more than the quality of the car. A study by Scott Hardman and Gil Tal of the University of California Davis showed that 20% of owners traded their electric car of the future for a gas-powered car of the past. The primary reason given is the lack of charging facilities, particularly level 2 or higher charging. Level 1 uses the standard 110V wall outlet, while Level 2 uses 240V (like a clothes dryer or stove), and Superchargers made by Tesla provide 480 Volts of charging.

According to the study, of those who switched, 70% lacked access to that Level 2 charging. Per *Bloomberg's* auto guru, Kevin Tynan, "if you don't have a level 2, it's almost impossible." Plugging the new Mustang Mach-E electric into a Level 1 charger for an hour, results in just three miles of range and just 36 miles overnight, while 3 minutes at a gas pump could give the gas-powered version 300 miles.[270] These owners were primarily younger and more likely to live in apartments or other shared living arrangements and thus less likely to be able to install permanent recharging systems.

With any technology, the early adopters push those solutions which will ultimately fit the defined need. Early adopters chose VHS over Betamax, Blueray over DVD-HD, and Windows over DOS. The fact that the early adopters of electric vehicles have chosen to revert back to old gas-powered technology means that the concept simply didn't solve the problems for which they purchased a car.

What happens to these vehicles down the road, especially if they are abandoned as not fulfilling the owner's needs? While they may be non-polluting on the roads, that isn't true at the end of the road. As Dr. Paul Anderson of the Birmingham Centre for Strategic Elements and Critical Materials notes, "In 10 to 15 years when large numbers are coming to the end of their life, it's going to be very important that we have a recycling industry. Currently, globally, it's very hard to get detailed figures for what percentage of lithium-ion batteries are recycled, but the value everyone quotes is about 5%".[271] With lithium supply likely to be extremely short when moving to grid storage, it will be critical that none of it gets wasted through non-recycled batteries. Few are aware of just how an EV battery is constructed. It isn't a single structure with 6 side by side cells, like a Lead-acid car battery. A Tesla battery is built with thousands of smaller batteries somewhat larger than an AA battery, wired together in small groups, and then the groups are wired together.

Not so easy to recycle, and only a few companies are currently set up to do so.

The world's waterways are currently crossed by more than 60,000 merchant ships of every kind, from small ferries that carry a couple of cars to large container ships carrying upwards of 24,000 twenty-four-foot trailer equivalents (up to 800,000 tons of cargo). To fully comprehend the magnitude of the shipping industry, check out marinetraffic.com, which provides a plot of all of those 60,000 ships' current locations as reported by their AIS transponders. Nearly all these vessels are powered by fuel oil, just one of those products refined from a barrel of oil. Without fuel oil, just how will these ships be powered?

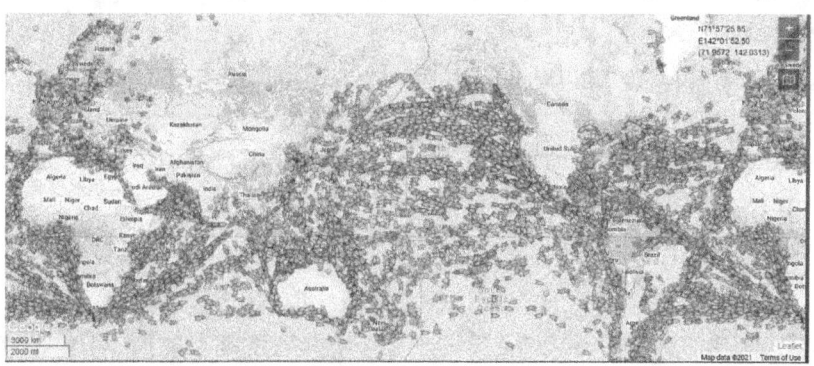

Figure 61 Real-Time AIS Tracking

Several proposals for ships powered by sails have been put forward. Swedish designers have been working on a design called Oceanbird, which would feature five large solid wing sails for when crossing the oceans but would still require an engine to maneuver when close to shore and when the wind is not blowing. The design is intended to replace 90% of the fossil fuel requirement.[272]

The French shipping line, Neoline, is designing 136-meter-long vessels capable of handling 500 cars and 280 containers and already have a contract with Michelin tires for the North American to Europe leg of the trip.

Other projects such as sailcargo.org (building an all-wooden schooner) and Grain de Sail's 72-foot schooner[273] can carry small loads of around 50 tons across the ocean under sail.

The Oceanbird and Neoline designers admit that journeys will take 50% longer, and "initially it might mainly be a way for companies to lower the average emissions of their fleet."[274] Not intended to actually move significant material but to allow them to advertise themselves as green.

Even with the current collection of 60,000 large ships, there is a backlog of shipping in California ports. To meet the future's shipping needs, 100 of these sailing ships will be needed for each of the large container ships that now ply the oceans. Further, the predictability and schedule of these ships will be at the whims of nature, and no longer will man be able to outwit the storms. Sailing ships can only go as the wind dictate, unlike the engine-powered vessels, which is why it took only a few years after the steam engine's invention for it to take over for sails.

Within the green strategy, rail is central to mass transportation over distance. Many look at Europe and Japan with their extensively developed train system, complete with high-speed rail to get from city to city. Two key factors make those systems good for them and less workable in the United States.

The U.S. geography is such that the distances from point to point are much more extreme. A few years back, my family and I took Eurostar high-speed rail from London to Paris during our vacation to Europe, a distance of 106 miles, a safe, fast, comfortable, and easy journey. However, leaving Washington, DC and going to the next nearest nation's capital city would be either the 525 miles to Ottawa, Canada, or the 2550 miles to Mexico City. A much more realistic comparison would be the construction of intrastate rail systems. And while it is possible to build those intrastate rails, the demand is not there. There have been programs to build Dallas-Houston and Los Angeles-San Francisco high-speed rails, but they fail from the local population's protests and when the lack of demand for such travel is revealed. Experience has shown in the United States that one of the main reluctances to using the train is that cities are so large that once disembarking from the station, the rider can still be far from his destination and must transfer to other means of transportation. Which leaves them thinking, why not just use the different means of transportation?

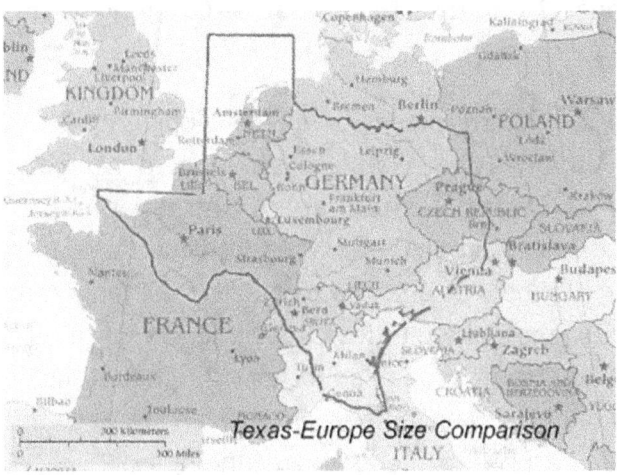

Figure 62 Europe as Compared to Texas

Until the lockdowns, modern society had grown accustomed to traveling, whether nationally or internationally. The **Green Solution** calls for the development of regional high-speed rail, but no provision has been given for more extended travel. The isolation of the United States relative to our trading partners in Asia and Europe begs the question of how that travel will happen under the new plans.

Fifteen years of development and research have gone into developing solar planes, and the result is less than could be helpful for most of the population. In 2013, *Solar Impulse,* a solar-powered aircraft did successfully cross the American continent. However, the plane with a wingspan the same as a Boeing 747 was capable of only carrying a single person and flew at speeds of less than 60 mph,[275] a speed at which it would have been just as easy to drive a car. Compare the Singapore Airlines routine flight that travels 9,500 miles in 18.5 hours with 300 passengers in an Airbus A350-900ALR to the *Solar Impulse* travel of San Francisco to Washington DC in just under 2 months, a rate that could be achieved with a Ford Model-T.

Several all-electric planes under development will handle flights under 500 miles for a couple of passengers, likely handling travel requirements for the elites but not capable of mass travel. Cessna was the first to enter the market with the Caravan model, which ultimately should be able to carry up to 9 passengers at 114 mph for a total distance of 100 miles,[276] although so far, it has been limited to flights with the crew. The future of flight will be determined by the technology development in power storage units. It won't, however, be a future of existing technologies.

There is no indication to date that travel will be possible on an international scale by the mass of travelers, so those pesky lockdowns with their unintended consequences of shutting down travel may become the new way of life in the pending dark future. Responding to the fearmongering of the doom and gloomers without regard to how life will be lived is the danger of supporting outright bans on fossil fuels.

11

The Dangers of the Green Solution

> *"...if you know your enemies and know yourself, you can win a hundred battles without a single loss. If you know only yourself, but not your opponent, you may win or may lose. If you know neither yourself nor your enemy, you will always endanger yourself."*
>
> Sun Tzu, **THE ART OF WAR**

As with any great revolution, there are dangers, and there will be winners and losers. The **Green Solution** developers seem to be intent on driving headlong into the project without understanding all the consequences of such a revolution. As Albert Einstein once said, "Three great forces rule the world: stupidity, fear and greed." Which of these will eventually be the driving force for our decisions?

The Battle

The rhetoric of the **Green Solution** pushes the idea that humanity will be competing for increasingly limited resources and that the winners in that competition will be endangering the lives of the losers. This may well pave the path towards conflicts, whether small and local or international in nature. If the world is looking at the collapse of ecosystems which results in driving mass migration of climate refugees which causes environmental and economic instability, then the world will be facing an unprecedented threat of conflict.

As each member of society is asked to take extreme measures and endure extreme sacrifices in the name of climate change, we face the issue that Mancur Olson referred to as "concentrated costs with dispersed benefits." In his 1965 work, *Logic of Collective Action,* Olson discusses that when any group attempts collective action for the common good, there is an incentive for individuals to obtain a "free ride" to gain the benefits without the sacrifice. In any scenario where progress is measured by millions of little steps over decades, the impact of any one decision can easily be dismissed.

In order to succeed, the battle against climate change will take money, but we must consider whether that money will be forthcoming. A 2018 Associated Press poll found that just 16% of the population would pay up to $100 per month to combat climate change, 23% up to $40, and 57% up to $1 per month. Overall, 43% were not willing to pay any money at all for climate change.[277]

Figure 63 Who is willing to pay to fight climate change[278]

This leaves governments as the only source of money for the battle. As long as those who are unwilling to pay their own money for

The Dangers of the Green Solution

climate change fail to see government money as theirs, the governments will be able to get away with spending on the projects. And significant sums of money will be needed as for each $1 spent on climate projects, those projects are expected to produce 11 cents worth of climate benefits.

Technology may change, the behavior of humans doesn't. While there has been a rise in national activism against fossil fuels, too many of those same people have become activists in local matters, although often in direct opposition to their national positions. These local activists, better known as NIMBYs (Not In My BackYard), object to and protest against any project that would affect them personally. If a project pushes their national agenda, they approve, as long as it remains someone else's problem. Many anti-wind NIMBY groups complain about the low-frequency sound effects, including sleep disturbance, headaches, dizziness, vertigo, nausea, and irritability.[279] As Tina FitzGerald, who lives on a Vermont farm near a proposed wind farm placement, said, "Anywhere I walk on this property, we'd be able to view them, and we'd be able to hear them... There should be a place for these -- someplace that isn't going to impact families quite so much."[280] People may love renewable power projects; they just don't love the one next door.

As residents of Martha's Vineyard, John Kerry and various members of the Kennedy family, as mentioned earlier, are classic examples of these personalities that while they are opposed to wind power being placed offshore near their home, they also invest heavily in wind companies, take governmental positions traveling the world to promote it and even sign international climate accords. Vineyard Wind, the offshore wind farm planned near Martha's Vineyard, may be able to start to install their first wind generator sometime in late 2021, after nearly 10 years of opposition.

It only takes a few local NIMBYs to influence a local government. In February 2019, about "five dozen" people testified and made their opinions known to the San Bernardino County board. The influence was enough to lead the board to vote 4-1 to ban utility-sized concentrated solar projects in the nearby deserts, despite the county's commitment to go carbon-free by 2040. The county's carbon-free plan was centered around those very same solar projects providing a majority of the daytime power.[281] Many boards find fixing the short-term discomfort of protesters in the meeting easier to handle than worrying about long-term problems. Problems that will present themselves long after they are out of office.

The opposition to these projects has been using social media to organize and spread information on that which they oppose. Their efforts at raising awareness and encouraging the public to testify at hearings mean that far too often, the opposition greatly outnumbers supporters when the topics are being discussed. The best a developer can hope for is expensive delays to a project, and the worse is the fact that a $500 Million wind farm project can die because 50 people complained to the county commissioners and caused a fuss at a commissioners meeting.

In 2009, BusinessWeek reported it "turns out there's something called the Starbucks Rule when it comes to siting wind farms. Plot where Starbucks are in the general area, and then make sure their projects are at least thirty miles away. Any closer and there's going be too many NIMBYs who would object."[282]

With the large number of power generation devices required in the move to a fossil fuel-free society, more and more projects will need to be closer and closer to where people live. It is then anticipated that the pushback from NIMBYs will create an increasing number of roadblocks to these projects. While these citizens may be approving of the projects, they may not approve of where they need to be located. Installation of those 500,000 to 750,000 wind turbines required for an "electrify everything" grid maybe, in the end, impossible to situate.

There will always be those for whom just protesting in the council chamber will never be enough. The radical practices termed eco-terrorism has been on the rise, a new way to protest. The latest battlefield for fossil fuels has become the pathways that get the raw fuels to the refineries and power plants. Recently, environmental terrorist groups have been busy targeting the railroads and causing derailments. In 2014, an average of one oil train derailed every five days. Two young women from Washington state were charged in December of 2020 with terroristic plots to damage BNSF trains by placing shunts on the tracks to disrupt the safety signaling. The "shunts" electrically short the two rails together and will cause the emergency brakes on the trains to engage, possibly causing damage to the couplings and potentially causing derailments of the separated cars. The filings claim that they carried out the plan "in solidarity with Native American tribes in Canada seeking to prevent the construction of an oil pipeline across British Columbia, and with the express goal of disrupting BNSF operations and supplies for the pipeline."[283] Other Antifa affiliated terrorists have succeeded in derailing an Amtrak train, also in Washington.[284]

The protest against the trans-British Columbian pipeline is ironic since that pipeline is being considered largely because of the halting of the north-south Keystone XL pipeline. A pipeline designed to ultimately deliver the same product from the Alberta tar sands to refineries in Texas. The protests against Keystone center around historic Native American lands and dangers to the whooping crane, greater sage-grouse, and the swift fox. There are just four choices when it comes to the Alberta product: a western pipeline, a southern pipeline, ship it by train, or none of the above. The activists are pushing for the latter.

If protesting alone is not enough, there have been other methods used to slow the use of pipelines. On May 8, 2021, malicious hackers named DarkSide used a ransomware attack on the Colonial Pipeline, which carries 10 million gallons of gasoline, diesel, jet fuel, and home heating oil from Texas to New Jersey, representing 45% of the fuel consumed on the East Coast. [285] While enough fuel is stored in the Northeast to handle two days of pipeline shutdown, the loss of the opportunity of shipping product through the pipeline can have long-term issues. The shutdown requires that the refineries either find places to store the 10 million gallons a day of extra production, when storage is already tight due to reduced usage during lockdowns, or shut down production, which causes cutback to those feedstocks used in petrochemical production.

National and International terrorist groups centered on the politics of the environment have sprung up in recent years, including the Individuals Towards the Savagery (ITS), Earth First, Earth Liberation Front (ELF), and Animal Liberation Front (ALF). Groups like the ITS are described as "disregarding human life and hold violent ideologies against progress and technologies." These groups "represent a blend of anarchists and apocalyptic, extremists whose philosophies are rooted in Marxism, socialism, feminism, postmodernism and Eastern religions, and who strived for the end of modern civilization." While these groups have launched around 200 attacks over the past couple of decades, they have not caused any fatalities to date, as a result, these attacks have been labeled as vandalism, sabotage, or trespassing. So to date, their actions haven't been labeled as terrorism, but that may well change in the future as the stresses from the political rhetoric of change increase. Due to conflicts over water, international threats continue to exist from groups like Boko Haram, Al-Qaeda in the Islamic Maghreb (AQIM), the Movement for Oneness and Jihad in West Africa (MOJWA), which are radicalizing their populations using environmental issues as a proxy for their ideological goals.

As the sense of injustice and desperation spread among those most affected by either climate change or the response to the change, the threats of terrorist action, whether by a lone wolf attack or coordinated groups, increase. We can be sure of one thing: the scale of the disruption being considered will cause a reaction, some of which will be violent.

The End of the Wonder Chemicals?

Those petrochemicals used to construct everything from buildings to electronics to fabric to medications will have to be replaced if the world is to be free from fossil fuels. The world can face two choices: replace the chemicals or move forward without them to a world that would look very much like a pre-industrialized society. Without them, the industries reliant upon those chemicals could no longer operate. All those benefits in health and safety would fade away. Life in society would lose all that progress that was made in the last 200 years.

If you need a transmission for your aging vehicle, it is more likely that another transmission could be acquired by scavenging a wrecked car from a junkyard rather than purchasing all the components and building it from scratch. Likewise, when we need the component feedstock chemicals to make petrochemicals, it is easier to extract them from the soup of compounds in crude oil and coal rather than build them from the component materials. The current feedstock industry uses large amounts of power, upwards of 20% of industrial energy consumption, to complete the chemical transformation into useable petrochemicals.

The first law of thermodynamics provides for the conservation of energy within the context of thermodynamic processes. This law essentially states that the amount of energy received by breaking a chemical is the same amount of energy required to create that chemical in the first place, once heating and other losses are taken into account. Further, in reality, 100% efficiency in any process can never be obtained.

Without fossil fuels, the chemicals will need to be obtained by other processes. The basic argument is that enough is known about chemical processes that fossil fuels are unnecessary, as all the required feedstock chemicals can be created from non-organic sources. In theory, CO_2 and water (after dividing the H_2 from the O) can be rearranged to produce natural gas (CH_4) by applying sufficient energy from renewable electricity. With enough power and access to enough elements, any of the necessary feedstocks can be produced in theory. There is no question that the idea is correct, but it comes down to costs and energy. Whether this scheme is affordable comes down to the cost of electricity. At 4 cents per kilowatt-hour, some of the simpler compounds become financially viable, while at 2 cents per kilowatt-hour, some of the more complex compounds become worthwhile undertakings.[286]

The Berkeley-based startup company, Opus12, was created in 2015 with the intent to develop the equipment and methods for the industrial creation of these chemicals. To date, they have done significant research but are yet to release any product to the market. The business plan for Opus12 is to convert emissions (primarily CO_2) into syngas and ethylene using washing machine-sized machinery. Once the concept is proven, similar machines could be developed for the other primary elements: propylene, butylenes, benzene, toluene, and xylenes. A future based on the production of petrochemicals from the Opus12 process would require the development of industrial-size operations.

There are three primary limitations to the plan to synthesize all petrochemicals: first, the energy cost. The average electricity cost in 2017 was 10.5 cents per kilowatt-hour. The estimate of renewables may get 15 cents to 30 cents per kilowatt-hour, depending on the assumptions. [287] [288] Some sources quote lower prices, but these often are wholesale prices which include government subsidies, make future projections, and suffer from unreasonable capacity factor calculations. Secondly, that enough surplus electricity will exist to perform these operations on a scale required to replace all the chemicals derived from petrochemicals and thirdly, that the processes themselves can be structured without the use of fossil fuels or rare earth metals so that the process is self-sustaining after the initial build. Scaling up the energy production to create these chemicals would require between 2 and 2.5 times the current electricity generation.

Several companies have introduced the use of biological processes to replace oil in the generation of petrochemicals. DuPont has introduced the concept of genetically modified bacteria as a source of deriving these chemicals. A unique fiber called Sorona has been available for several years and is currently used in producing carpets from Mohawk and Karastan. The bio-fiber uses less power and water, which DuPont claims is a 56% reduction in greenhouse gases. Other products from biosynthesis include Biolsoprene and BioAcrylic, both fibers are in production in small quantities. As biotechnologist Dr. Craig Venter states that "the goal is to replace the entire petrochemical industry"[289] with synthetic DNA. The idea is that in the end, they will be able to "Use synthetic biology, a promising new technology that lets scientists reengineer the genetics of living organisms, to take on the fossil-fuel industry— and do the whole thing with pond scum."[290] The question remains whether the anti-GMO protesters will allow the generation of literally thousands of new organisms to provide the material or whether, like the NIMBYs blocking wind, they will stop this.

Shortages Abound

Problems in producing those base chemicals and feedstocks will have but a single end result, shortages. Manufacturers will necessarily cut back on production due to troubles acquiring the raw materials they will need. Shortages of plastics, medicines, fabrics, and even foodstuffs, could result in all sorts of unexpected results.

As the San Francisco Federal Reserve Bank notes, "Rapid wage increases or rising raw material prices are common causes of [cost-push] inflation." The inflation of the 1970s, due to rapid increases in the cost of energy, is an example of this style of inflation. As opposed to demand-pull inflation, cost-push inflation stems from when production costs increase, forcing the producers to raise prices that eventually make their way to the consumer in the store. Since the world is currently flooded in currency, we already face inflationary pressures with too many dollars chasing too few products. The raw material shortfalls may cause even fewer products to be available, potentially triggering hyperinflation. The hyperinflation of the Weimar Republic in Germany during 1922-1923 was in part caused by the devaluation of the German mark, causing raw materials to become overpriced, and few nations wanted to trade their products for questionable marks. Germany's shortfall of raw materials started its issues rolling. Will we do the same?

As soon as the government then steps in, what happens is best described by Thomas Sowell as "price controls almost invariably lead to declines in the quantity and quality of what is supplied, to hoarding and to black markets – whether the price that is being controlled is that of food, housing, gasoline, medical service or other goods and services."[291]

As a lesson on the interconnectedness of our supply chains, the recent semiconductor shortages have impacted everything from automobiles to cell phones. Rabobank's Global Economics & Markets desk commented, "technological wonders of a global economy based on just-in-time supplies of a few key inputs from only a few locations, and then demand surged due to a virus that ran rampant through said global economy, and supply chains got snarled for that, and other reasons; and now a lack of silicon chips even impacts on the price of potato chips (in the U.S.) and chips (in the UK."

Leaving the Majority Behind

White's Law: culture evolves as the amount of energy harnessed per capita per year is increased, or as the efficiency of the instrumental means of putting the energy to work is increased.[292]

Several circumstances primarily out of the control of individual nations determined which countries became the advanced, industrialized nations and those that are still developing. The developed nations exist in temperate climates, as some nations are too hot, such that the daytime heat interferes with the worker's productivity, while others are too cold, requiring too much effort to remain warm. Variations in soil, water sources, and altitude all affect the ability of nations to develop. The nation's natural resources are less important than other factors, as evidenced by the fact that Congo with substantial natural resources is among the poorest, while Japan, with limited natural resources, is among the richest. Resources can help a nation succeed but are not enough in themselves, but energy plays a bigger role.

Across the globe, 2.5 Billion people use wood, biofuels, or animal waste for their cooking and heating needs. An estimated 1.8 billion people lack access to adequate clean water. While an estimated 1.1 billion people practice open defecation. Currently, 47 nations are considered undeveloped[293] as listed by the UN in their Least Developed Countries (LDC) category and another 60 nations are deemed underdeveloped. It is said that "there are billions of people in underdeveloped countries who are currently living in the low energy economy medieval days that developed countries left behind a century ago... They have yet to join the industrial revolution, and without oil and natural gas, they may never get that opportunity."[294]

The developed world represents 1.3 billion people or roughly 17% of the world's population.[295] The projections are that as much as 50% of the world's population could join the developed countries listing by 2050. But this can only happen if they have sufficient energy available to them.

Every industrialized nation has made the transition from non-industrialized to its current state on the back of fossil fuels. They did this because there is no other high-density energy source that gives man the capacity to do more work than required to sustain him.

The IPCC encourages these underdeveloped nations to leapfrog to renewables. The idea is that these nations can move from the historical usage of wood and natural sources of energy directly to renewables while leapfrogging the intermediate step of fossil fuels. My friend Cedrick's country of Cameroon has been given the choice of solving its problem of the lack of energy in the rural areas by either buying solar panels from China or powerplants from France. Providing only high-tech solutions to solve issues in low-tech countries is placing roadblocks in their ability to join the developed world.

John Briscoe wrote, "Time and time again I have seen NGOs and politicians in rich countries advocate that the poor follow a path that they, the rich, never have followed nor are willing to follow."[296]

Will these nations be willing to go along with this program? The developed nations are taking advantage of them. As Stein and Royal said, "Understandably, it is hard to imagine the billions of people in underdeveloped countries who have yet to experience anything like the industrial revolution and who are surviving without any of the advantage's fossil fuels are providing to the lifestyles of those in developed countries."[297] The Paris Climate Accord hopes that these nations are willing to take "climate finance" dollars to industrialize as the developed nations dictate rather than undertaking their own industrial development. As such, third-world countries become the biggest losers as the world moves to a fossil-free existence. Those who live in the lesser developed nations face a choice, face life with no hope of rapid development or consider migration to the developed world. The engineering world is full of young, bright, talented individuals who have left all sorts of countries, China, India, Bangladesh, Pakistan, and even Cameroon, to find better opportunities in the developed world where access to the benefits of fossil fuels makes more things possible. If the result of the environmental movement means leaving the majority of the world's population behind, the developed nations will soon feel increased immigration pressures as the ambitious of all nations attempt to reach them.

Population Control

In the 2018 film, Avengers: Infinity War, the plot revolves around the hunt for five infinity stones. The villain Thanos collects the five infinity stones to create a "universe free of suffering"[298] by eliminating half of the universe's population. A bit extreme, but was this wrong? Many think not. In fact, there is a Reddit thread named "ThanosDidNothingWrong."

The Canadian Broadcast Company posted to their website on October 25, 2019, a troubling article in which it said, "The argument is that if there were fewer people on Earth, greenhouse gases would be reduced and climate change could be averted." As Sir David Attenborough has said, "Instead of controlling the environment for the benefit of the population, perhaps we should control the population to ensure the survival of the environment."

Those over-excited books about population control in the '70s and '80s were correct in their concept that the world's population would explode. Much of their reasoning was based on Thomas Malthus's theories, so it was concerned that "Population, when unchecked, goes on doubling itself every 25 years or increases in a geometrical ratio," a rate which, like so many models, hasn't reflected reality. It was thought then that food would be the limiting factor that would reduce the rate of increase. The green revolution of Dr. Borlaug and fossil fuels fed the growing population, so what would be the next controlling factor. As countries develop and the girls and women of the society receive more education and employment opportunities, the birth rate declines considerably, sometimes to less than the replacement birth rate. The population is expected to decline in the second half of the 21st Century because of that.

Still, there are those who would prefer to see a significantly smaller population than currently exists. In 1982, the National Institutes of Health published an article by Yanjiu Renkou in which he argued, "Therefore, a well-planned program for population control is essential for achieving decent quality of life."[299] While more recently, the Center for Biological Diversity has stated that, "We can reduce our own population and consumption to an ecologically sustainable level in ways that promote human rights; decrease poverty and overcrowding; raise our standard of living; and allow plants, animals, and ecosystems to thrive."[300]

After all the overpopulation hype died down, the term "population control" has fallen out of favor. Instead, the same concepts have been relabeled into rights-based and social justice language intended to make it more palatable, namely reproduction justice. As USAID wrote on their blog for World Population Day, "people, planet, prosperity, peace, and partnership. By slowing rapid population growth, family planning can help to decrease the sheer number of poor people." A plan to reduce poverty by eliminating births rather than providing them with the essential energy that could do the same job.

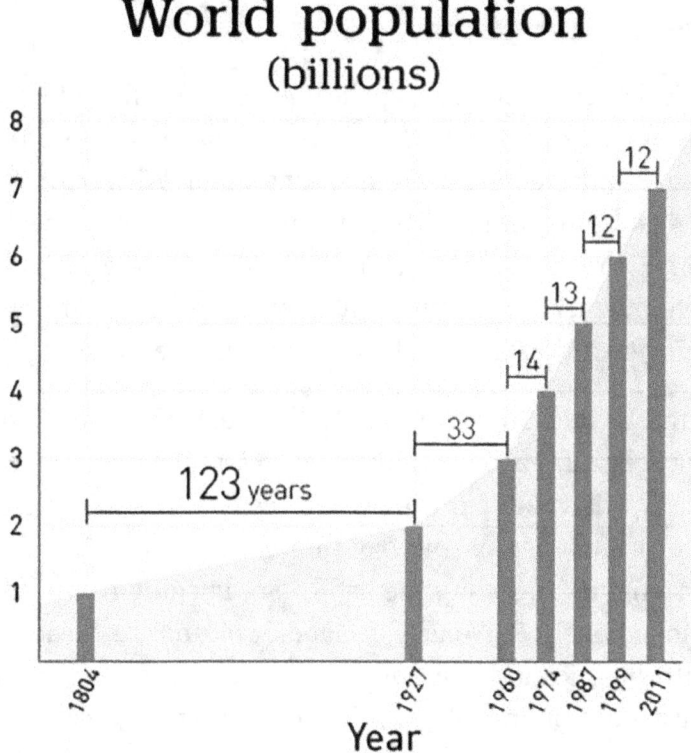

Figure 64 World Population Growth

The argument is that human population growth is at the fundamental core of the issues around energy, resulting in more people with limited energy, meaning less per person, therefore should we not take some action regarding population? When organizations like the Center for Biological Diversity run articles on Reproductive Justice, it should be noted that this isn't about the right of women to control their reproduction but instead is designed to limit the population in the name of the environment.

It is true that when the Black Death wiped out 25% of the European population in the Middle Ages, the result was improved lives for those that remained, but who gets to choose? Thankfully, those who speak of the "final ten years" seem to have plenty of ideas on that.

Exposure to Natural Events

Geographical circumstances weigh heavily on the ability of nations to gain developed status, with the lesser developed nations often in danger of natural events occurring around them. With our current energy industry providing reliable, plentiful, and cheap energy at our demand, those in the developed world have largely become unaffected by many natural events, but they may affect the leapfrogging countries in their dependence on renewables.

Between the 14th and 20th of April 2010, the volcano Eyjafjallajökull erupted in Iceland, expelling a giant ash cloud covering much of Europe. For six days, 20 countries were forced to close their airspace to air travel due to the problems of visibility and the possible danger to aircraft engines ingesting the ash.[301] So, imagine the effect on exposed solar panels. Previously, Icelandic volcanos have cause ash clouds with significant disruptions such as "the seven-month-long Laki fissure eruption in 1783, which caused a famine that claimed 20 percent of Iceland's population and lowered Northern Hemisphere temperatures by an estimated 1°C."[302] Icelandic volcanos are expected to erupt every 5-10 years on average. Even during the writing of this book, yet another Icelandic volcano, Fagradalsfjall, has erupted, though luckily without significant ash production, and National Geographic predicts that it foretells of a decade worth of activity. Of course, there are those 10 years again.

While the United States, outside of Hawaii, is generally not exposed to volcanos, Mount St. Helens' eruption proves it is merely improbable, not impossible. With Mount St. Helens, the ash-covered some 22,000 square miles.[303] During May and June of 1980, any renewable energy sources within Washington and Idaho would have been damaged.

Eyjafjallajökull and Mount St. Helens both affected the regions directly downwind from the volcanos, a regional disaster for which other regions could make up the power loss. However, consider the 1815 eruption of Mount Tambora in Indonesia, the largest volcanic explosion known to man.[304] The year is regarded as the "year without a summer" due to the amount of ash and soot put into the atmosphere, which was enough to drop the worldwide temperature by 3°C. The lack of sunshine affected crops worldwide, and such an effect on solar panels could only be imagined.

The world is full of natural events that would expose solar and wind power sources to reduced performance and, at worst, significant damage. The effect of items landing on solar panels or in the air over the panels is obvious. However, a 2001 study by researchers found "up to 50% loss of wind power capacity caused by insect deposits on rotor blades."[305] This study was done on relatively small and light insects like butterflies. Imagine the impact of a swarm of heavier and dense locust, such as moved through Asia and Africa in 2019 and 2020.

In June 2020, a Saharan dust cloud[306] reached the United States bringing pretty sunrises and sunsets but, more importantly, diffusing the light that would fall on the solar panels. A German study of the Saharan Dust clouds that commonly reach Europe stated that "When it comes to Saharan dust outbreaks, the photovoltaic output is reduced not only through a significant increase in atmospheric aerosol content by 10 to 20 percent, but also through dust deposition on the photovoltaic modules on subsequent days."[307] The research noted that there is a lesson to be learned that whenever rain leaves mud deposits on cars, it leaves it on the solar panels.

The effect of wildfires on solar generation has been well documented. During 2020, California experienced about 8,000 wildfires of various sizes. In September, a fire in Sonoma county occurred upwind from the Sonoma Valley Unified School District, which uses solar panels to supplement the power from PG&E. The school district reported a 58% drop from the week earlier and a 66% reduction in solar energy production on their active solar PV system.[308] The effect is that school days were lost, and Sonoma Clean Power has created a program to help the school install battery storage systems to overcome the loss of power from future fires.[309]

Among the fatalistic descriptions of the future earth as an effect of climate change is often an increase in dust and sandstorms. The UN Convention to Combat Desertification (UNCCD) has several initiatives design to detect and combat this problem. According to some of their models, while the United States has not seen dust bowls since the 1930s, the UN predicts them for 2030. The National Renewable Energy Laboratory released a study in June of 1992[310] describing the effects of sandstorms on photovoltaic solar panels and expressed concerns with scratching the surface, the soiling of the surface, and wind loading structures. The report notes that in the western deserts, where the large solar farms are constructed, severe sandstorms average about twice per year. To minimize damage from these sandstorms, PV arrays would need to be movable (horizontal during the storm), at least 6 feet off the ground (to avoid the largest and most damaging sand particles) and have a plastic scratch-resistant cover.

Further, to maintain peak performance, most PV solar panels need to be washed at least twice per year and whenever the panel is coated due to wildfires, volcanoes, dust, or even pollen storms. Leasing and warranty issues can complicate the who and when such panels are cleaned.

Likewise, these sandstorms will be no friend of the wind turbine. Studies show that the sand-blasting effect on the blades during a sandstorm can lead to a loss of performance, causing a 40% power output drop from clean blades and a 25% reduction in annual energy output. Since the blades cost around 20% of the entire structure, replacing them after storms could become quite expensive. When GE installed a 13-turbine wind farm for the country of Oman, they were required to design unique sand-proof versions that could filter out the sand and protect the internal workings from the effects of sand while still trying to limit the internal heat buildup.[311] All these modifications reduced the total potential energy output of the turbine while increasing the expense.

The idea that renewables generate power when the sun shines and the wind blows and sit idly by when they don't is overly simplistic. The poorest nations, who are being asked to leapfrog, are in the sub-Saharan region where sand and insects are likely to damage the equipment. Will they standby and allow the developed nations to dictate to them?

Military Weakness

> *"Both WW I and II were won by the Allies, as they had more oil than the Axis Powers of Germany, Italy, and Japan to operate their military equipment and move troops and supplies around the world."*[312]

"War is the continuation of politics by other means." – *On War*, Carl von Clausewitz, 1832

For as long as there will be politics, there will be war. The last war fought before the introduction of fossil fuels onto the battlefield was the early part of the First World War, a battle famous for its brutality and slaughter of the trenches. It was a time where men buried themselves into the ground to avoid the spray of machine guns and other modern offensive weapons against which they had no defense. As horrible as those trenches were, they brought safety.

Sir Hew Strachan has written, "Trenches saved lives. To speak of the horror of the trenches is to substitute hyperbole for common sense: the war would have been far more horrific if there had been no trenches. They protected flesh and blood from the worst effects of the firepower revolution of the late 19th century."[313]

The advancement of technology in the last portion of the 19th century led to the development of machine guns, accurate artillery, and high explosives. A war that began with marching soldiers, horse-drawn support wagons, and battle tactics from the previous two centuries ended with armored cars, tanks, and trucks moving both soldiers and supplies.

From the outset, the strategy of World War II evolved around the restriction or capture of fossil fuel supplies, from the oil embargos against Japan leading to Pearl Harbor to the German invasion of Russia in hopes of capturing the oil supplies of the Caucasus. In August of 1943, the United States Army Air Forces executed the air raid on the Ploiesti Oil wells and refineries in Romania to restrict Nazi access to Romanian oil fields. The battle for the oil was important enough to endure the costliest (percentage-wise) raid of the whole war.

War in the 21st century is even more tied to the available oil and fossil fuel products to sustain a fight. During this century, the U.S. military has used between 77 and 80 percent of all U.S. government energy consumption.[314] In the modern era, war is defined by speed and surprise in its operations, faster vehicles, planes, and ships are always in demand. Every tool of warfare now needs fossil fuels to operate - tanks, armored personnel carriers, helicopters, fighter/bomber jets, drones, ships, submarines, and satellites. The end to fossil fuels means an end to the means of waging war, which is one thing if all parties participate but comes up short when one party unilaterally reduces the use of these tools. If there are no fossil fuels on the battlefield, expect to return to the brutality of the Great War.

Take the lesson of America's main battle tank, the M1A1 Abrams, as an example. Because of fuel demand in a battle situation, the tank was designed to burn any fuel that burns, including diesel, gasoline, even jet fuel. The tank's fuel capacity sits between 450 to 500 gallons (depending on the production run). The 60-ton vehicle typically gets somewhere around 0.6 miles to the gallon, so the 500 gallons gives it a range of about 300 miles.[315] While electrical power tanks have been discussed[316], the problems of providing electrical recharging on the battlefield are orders of magnitude harder than sending some guys out with Jerry cans of fuel.

To maintain the nation's security means maintaining the military fighting effectiveness. Until advances in technology can lead to a new form of motive energy or alternative means to fight a war, fossil fuels will need to be a mainstay in the military's arsenal.

Further, it is not just the fuel to move the military equipment that is at risk but the ability to construct the armored equipment in the first place. Companies like Clifton Steel[317], CMC[318], SSAB[319], and Leeco Steel[320] run small, specialized steel mills to create the plates for body armor, vehicles, and naval vessels. It is critical for the nation's defense that these fabricators be able to continue to use materials and energy at a profitable cost so that our soldiers, sailors, and airmen can do the job of maintaining the integrity of the United States. Besides steel, the military uses weatherproofing, plastics, medical supplies, and communications equipment, all constructed from fossil fuels or derivatives.

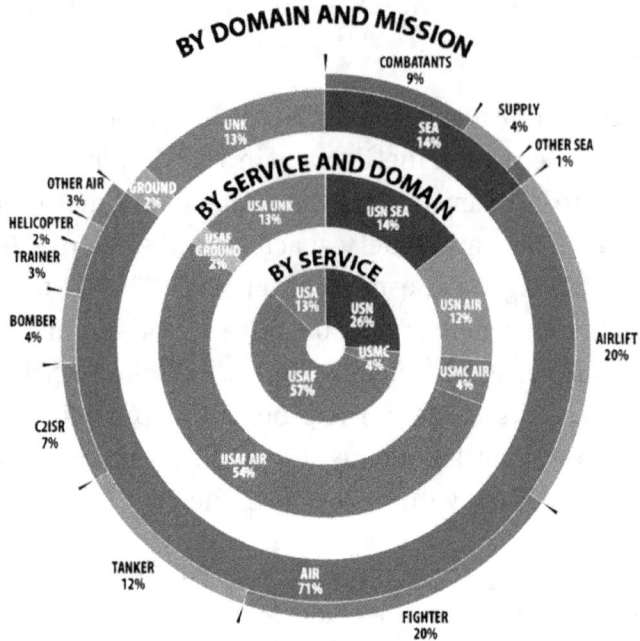

Figure 65 Use of Fuels within the U.S. Military

Roughly 75% of all the American armed forces' energy is in the form of aviation fuel. This is true for all six military branches, with buildings being the second-largest energy user, followed by transportation. Since the decommissioning of its autonomous power generation capacity over the past two decades in the name of cost savings, the Department of Defense relies upon the U.S. power grid for 99% of its electrical power. While the military does have limited local generator capacity, any power grid outage affects it as much as it does the rest of the public.

The proposed elimination of the sine qua non of oil production (gasoline) places at risk the ability of the military to access this needed aviation fuel. Overall, aviation fuels represent only 10% of a barrel of oil. Without a market for the other 90%, which the military cannot fully consume, commercial operations have no reason to create this fuel. Shortages of aviation fuel could put

America's and the world's security at risk as there are those who would take advantage of such a situation.

So, these are the possibilities, but what is the actual position of the military on the idea of using it to fight climate change? On September 11, 2020, the Secretary of the Army issued Army Directive 2020-08 ("U.S. Army Installation Policy to Address Threats Caused by Changing Climate and Extreme Weather"). As a result of the directive, The Army War College report, *Implications of Climate Change for the U.S. Army*, warns that "this includes national security challenges associated with or worsened by climate change, and organizational challenges arising from climate change related issues in the domestic environment."[321] The report deems that stewardship is an issue with the Army's institution declaring, "Army leadership must create a culture of environmental consciousness, stay ahead of societal demands for environmental stewardship and serve as a leader for the nation or it risks endangering the broad support it now enjoys." Fearing that "lagging behind public and political demands for energy efficiency and minimal environmental footprint will significantly hamstring the Department's efforts to face national security challenges." The Army feels it must keep current with public demands to maintain its positive public image else it "will impact the military's ability to receive the required funding." The public's view on climate change influences government policy, and that policy affects what the military can do.

The report addresses the need to create battle plans to counter the listed threats as a result of climate change which include:

1) soldiers: dehydration, insect-borne diseases
2) fighting conditions: changing geography due to rising seas and growing deserts
3) future engagements: food insecurity, power outages, extreme weather

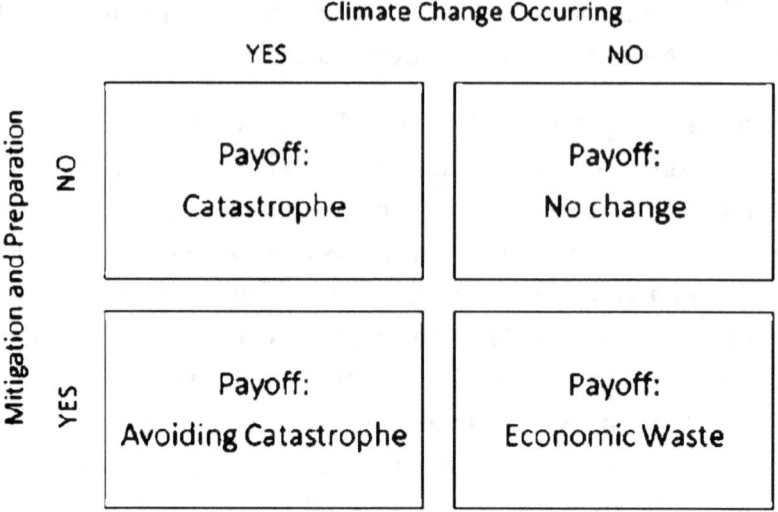

Figure 66 U.S. Army Climate Change Risk Assessment

The Army estimates that within 10 years, it must be prepared for disruptions to its fuel supply, which it plans to meet by "significantly increasing investment in more realistic simulation that incorporates the advances in virtual and augmented reality, [while] continuing to invest in the development of lower CO_2 emissions platforms and systems." Changes to the design to its equipment will be necessary because "its weapons systems and vehicles are not designed for energy and fuel efficiency or to minimize the impact to the environment."

Under the Biden Administration, the ground forces have been asked to cut back on the number of field training exercises, naval vessels are cutting back on time at sea, and the aviation arms have been instructed to cut back or save fuel through a reduction in the number of crewmen onboard aircraft, reductions in armaments being carried, and removal of "unnecessary" equipment. At any point, any of these changes could endanger the crews or the country.

Fighting the Oligarchs

When a gasoline power vehicle runs empty and needs to be refueled, the owner has choices, roughly 9,000 drillers[322] and 1700 tanker ships[323] supply crude oil to 135 refineries[324] that produce fuel to be delivered by 100,000 fuel trucks[325] to the 168,000 fuel stations[326] nationwide. The automobile owner has endless choices, and the market forces continually encourage newer, better service from some stations while eliminating those with older, inferior service. One of the valuable aspects of gasoline is that, for the most part, it is entirely fungible. Gasoline from many stations can be mixed and matched, and would still be perfectly usable. No gasoline is genuinely significantly different from any other gasoline, the minuscule additives notwithstanding. Of course, that is if gas stations are allowed to exist in the face of the Coalition Opposing New Gas Stations (CONGAS) which is pushing to ban new gas stations within Sonoma County, California, and limit the changes or additions to existing stations. Along with other groups such as Seattle-based Coltura and Stand.earth, they feel that "existing gas stations are providing all the gas currently needed" and aim to "accelerate the shift to electric vehicles."[327]

The electric vehicle turns that plethora of choices on its head. Now, the supplier of power is limited to the utility company. Within Texas, 70 companies are authorized to sell electrical power from the grid; most of these do not actually generate power but merely serve as the intermediary between the power generation sites and the consumer. Despite the numerous power generation sites and power sales forces, the Texas electrical grid is controlled and managed by a single entity, the Electrical Reliability Council of Texas (ERCOT), as part of Texas's Public Utility Commission. There are five Transmission and Distribution Service Providers (TDSP) within Texas who actually owns the wires, cables, and distribution network of the electrical grid for particular areas. Each of the five companies covers different regions without overlap. So, while the power may come from many places and be sold by several companies, the EV owner plugging his automobile into the grid is limited to just one TDSP and one electrical grid provider (ERCOT).

When trouble happens, it is easier to work with your local fuel provider. If satisfaction is not obtained, your business can be taken elsewhere. But as physicist Amory Lovins wrote in his 1976 Foreign Affairs essay, "In an electrical world, your lifeline comes not from an understandable neighborhood technology run by people you know who are at your own social level, but rather from an alien, remote, and perhaps humiliatingly uncontrollable technology run by faraway, bureaucratized, technical elite who have probably never heard of you."[328]

Denmark's government is looking to localize all their electrical power into a centralized wind power island with a planned opening of 2033.[329] In a classic case of putting all your eggs in one basket, the concept would provide all the electrical power for Denmark through offshore wind turbines all connected to a single artificial island 50 miles offshore, which would then route the power to Denmark's mainland and numerous islands, plus to other national customers. Here the entire nation would be at the mercy of a single entity that would control the island. Control the island, control the country.

The World Economic Forum (WEF), among others, has been increasingly calling for development to be undertaken by private-public partnerships with exclusive rights granted to the individuals and companies involved. The large, multinational corporations end up having immense power concerning the businesses in which they are involved. The question becomes what happens when the corporations operating as essentially oligarchs in their sphere of influence encounter cancel culture. The world runs the risk of having solutions and ideas canceled through the oligarchs' power in favor of an existing but less effective solution.

The push to electrify all things and force everyone onto the oligarch's electrical grid begins by banning any alternative to the preferred actions. Cities across the nation are pushing new building standards to push future homeowners in the direction of being fossil fuel-free. In 40 municipalities in California, natural gas for gas-powered stoves, water heaters, clothes dryers, and gas-powered furnaces have been banned in new construction. Adding office developments and restaurants to the ban completes the picture. As soon as enough locations prohibit gas use, the appliances that use that fuel will quickly disappear from the market, leaving everyone with no choices.

The new International Well Building Institute approval standard represents the latest oligarchy to reach out to control energy use when it comes to public buildings. In their standard, what energy sources a building uses contributes to the building's ability to receive approval and the right to post signage on their doorways. Signage that the population is being told to look for before entering a building for their own "safety."

If all these dangers exist when moving to alternative energy sources, one must ask whether energy is really the topic of discussion anyway. Christina Figueres, executive secretary of the UN's Framework convention on Climate Change, has stated that "This is the first time in the history of mankind that we are setting ourselves the task of intentionally, within a defined period of time, to change the economic development model that has been reigning for at least 150 years, since the Industrial Revolution."[330] For her, the goal isn't so much the energy as the economic model under which the world operates. The only economic model which has produced the wealth and prosperity that the developed world enjoys today is capitalism. According to Figueres, moving away from capitalism "is probably the most difficult task we have ever given ourselves, which is to intentionally transform the economic development model for the first time in human history." Saikat Chakrabarti, Representative Ocasio-Cortez's chief of staff and principal author of the Green New Deal admits that the bill wasn't about energy when he commented, "Do you guys think of it as a climate thing?" Because we really think of it as a how-do-you-change-the-entire-economy thing." Actually, it was designed to "advance social, economic, racial, regional and gender-based justice and equality and cooperative and public ownership." So, is the rise of oligarchs that would arise in such an economic environment be a problem or an intentional result?

In his paper, "Fossil Fuels improve the planet," Alex Epstein points out that "Renewable or sustainable implies that the ideal life trajectory is one of repetition, using the same methods or materials over and over. But that is an ideal fit for an animal, not a human being."[331]

The implementation of the **Green Solution** places significant dangers to those who proceed with the ideas. There are dark days ahead for those who jump in too fast. But is there hope?

12

Is the Future Dark or Bright?

Angels in the Bible always start their interactions with the command, "Do not fear." This is a wise command because the natural reaction to fear is to freeze and be paralyzed, two actions that do not solve problems. If the future is to be bright, then we must not fear, we must go boldly forth into the future and solve the problems that lie before us. We must, as in the Star Trek motto, "go boldly where no man has gone before."

It is not true that modern society takes a safe climate and turns it into something dangerous. Rather, the world was always dangerous, and in fact, the industrialized culture makes the world safe. It is the high-energy civilizations where the climate is most livable and life is the safest. [332] Nothing has done more to save lives and protect the environment than fossil fuels, along with the materials, technology, and industry that they have brought to humanity. Not everyone is fully aware of these facts, but you can now count yourself as more knowledgeable than most, having gotten this far in this book. So, we stand at the crossroads, do we go towards a higher energy civilization or a lower energy one? Which path will be cleaner? Which is more sustainable? Which reflects the best choices for humanity, which will lead to the utopian world that we all wish for our prodigy and which lead to the dystopian world we all fear.

As MIT climate scientist Kerry Emanuel said, "I don't have much patience for the apocalypse criers. I don't think it is helpful to describe it as an apocalypse."[333] It isn't time to cry apocalypse but take action and make decisions that will truly solve the problem. To not just apply yet another patch, hoping that by de-energizing the earth, all will be better.

Politicians from Joe Biden to Alexandria Ocasio-Cortez have spoken of the need to address climate change as "our time's Moonshot" or "our time's World War." These two analogies are poor examples to use as they misunderstand the past, as well as the future. In comparison to the moonshot of the 1960s, the risks to society differ widely. With the Apollo moon program, President Nixon's greatest fear was that two astronauts would be stuck on the moon and unable to return. As much as it would have been a tragedy, it is likely that the program's failure would have resulted in the deaths of a handful of volunteer astronauts who had signed up to take the risk. Indeed, we lost three astronauts in the Apollo 1 fire and 14 additional crewmen in the shuttle disasters, yet the space program continues. In the case of the comparison to a world war, the time frame involved is all wrong. The United States fought in World War I for 19 months and World War II for 46 months.

The **Green Solution** involves everyone in the country and possibly on the planet. Everyone everywhere is at risk of potential failure. The project to convert the nation away from fossil fuels is designed to happen over 30 years (360 months) and will continue on forever due to the required maintenance and updates to the solution. Failure of the program means the loss of the reliable, plentiful, and cheap power that all have enjoyed and, in so many ways, has driven the economy to great heights. No, this isn't something as simple as a World War or a Moonshot, but a far more complicated, risky, and expensive project.

Is the risk worth taking? Society is considering risking everything on an idea for which there is no consensus despite what they would have you believe. Over the last 500 million years, the earth's estimated temperature has swung widely over nearly 40°F, far more than the up to 4°F possible from climate change. Al Gore has repeatedly predicted the world would be polar ice cap free, sometimes imminently. However, as the next plot shows, while the world has been without polar caps before, we are far from that point right now.

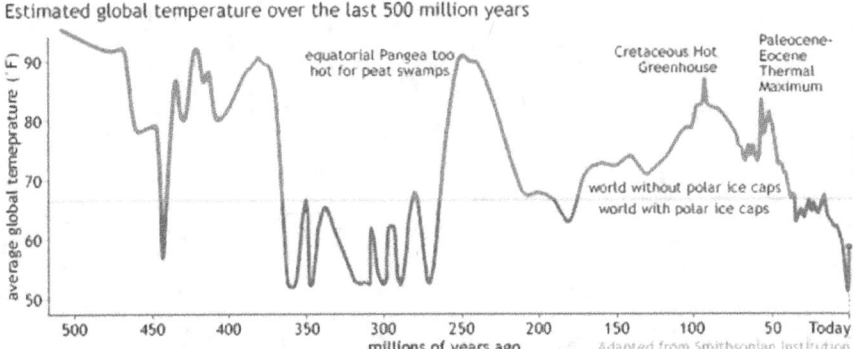

Figure 67 History Global Temperature Track

The temperature measurements over more recent timeframes are divided into two regions, one which determines temperature indirectly by things like tree rings and ice cores, and a second region where direct measurements were made, primarily due to the invention of an accurate thermometer. In both cases, direct and indirect, adjustments are required to remove variations that are not temperature related, such as changes in the measurement sites, whether a relocation of the site or physical changes like buildings or concrete nearby. Previously, the issues of computer modeling have been addressed, and the problem here is similar. Without specific ways to correlate, it is difficult to know where the adjustments are correct or not.

The question is: are we ready to risk everything for everyone over a long period of time over variations that may have modeling or correlation errors? And which doesn't represent the longer term trend of the planet's temperatures?

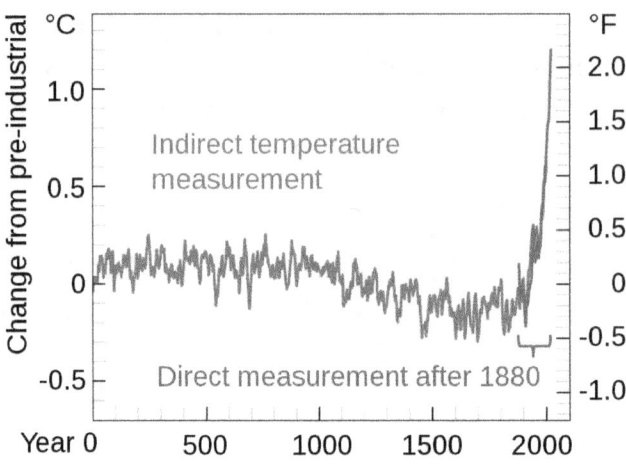

Figure 68 Global Temperature in the Common Era

Way back in the section on confirmation bias, a graph was presented showing four interpretations of the same NASA – Hanson data set. There are four lines because adjustments have been made to the interpretations, as measurements sites move or have changed conditions, the past is adjusted to, in theory, match the present. Figure 69 shows the raw and adjusted data for one of those data sets, and it indicates that cooling adjustments were made to the past, which implies warming in the present.

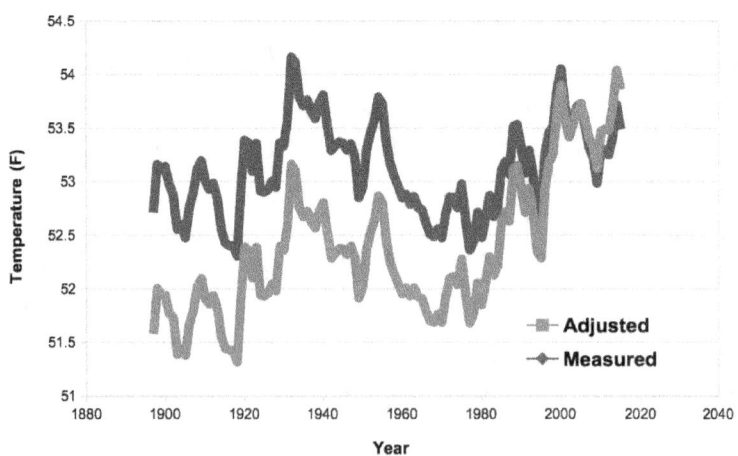

Figure 69 NOAA Average Annual Temperature (1218 Stations)[334]

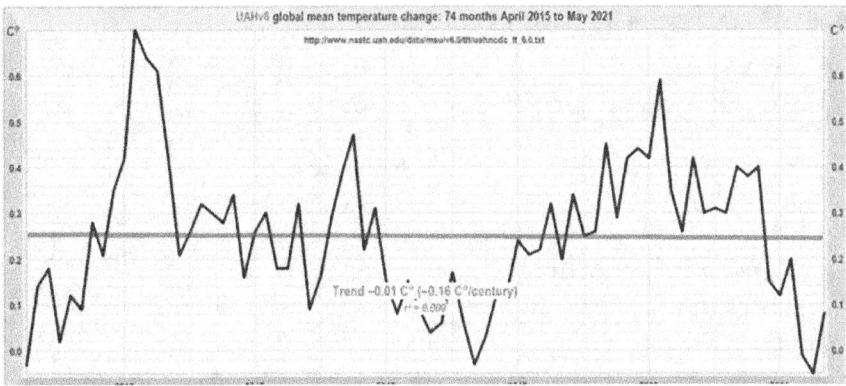

Figure 70 Global Mean Temperature Change April 2015 to May 2021 [335]

Without adjustments, the mean temperature just doesn't change (see figure 70). It might not have been colder in the past, it was just adjusted that way.

An Energy Transition?

Energy transitions happen in a society when certain factors reach their limits. As long as the current resources appear plentiful and reasonably priced, moving away from the status quo is difficult.

Even if great inventions come along, it is likely that the inertia of society and the actions of key elements of the economy will keep the status quo in place, although expanding with the new technologies. On the other hand, if the resources appear to be overpriced or at risk of running to exhaustion without an influx of new ideas, resources, and methods, then this can lead to systematic collapse. However, if new ideas, resources, or methods arrive and society adopts them, society is ready to transition to a new energy regime.

Table 15 Framework for Energy Transitions

		Future Resources and Opportunities	
		No New Ideas or Resources	New Ideas seem promising
Current Resources and Opportunities	Still Intact	Status Quo	Status Quo with Expansion
	Threatened or Exhausted	System Collapse	Energy Transition

In which place are we now standing? Some writers say, "We think some structural symptoms of such a transition are already in sight."[336] However, in their paper, Fischer-Kowalski and Harberl, also point out that more may be needed to make the change happen. As they state, "it will take a major 'push' from the part of nature, maybe even a manifest threat of collapse to accelerate a transition."

History has shown that it is often short-sighted to view the supply of any material as possibly running a risk of becoming threatened or exhausted. Frequently, they are only looking at the limited view of currently published numbers. However, the amount of material available at any one time represents the quantity needed at the moment. There is no value in a company or individual investing time and money to search out and find new supplies of resources when there is plenty already in the pipeline. The U.S. oil reserves in 1980 were estimated at 30 billion barrels. Over the next 33 years, the American oil industry produced 80 billion barrels, and the remaining reserves still sat at 30 billion.[337] Why go looking for more when there is plenty at hand, but when the outlook looks like a shortage may come, it is worth the investors' dollars to look for more. So maybe the status quo will be with us for a while longer, thanks to the improvements afforded by the latest in inventions.

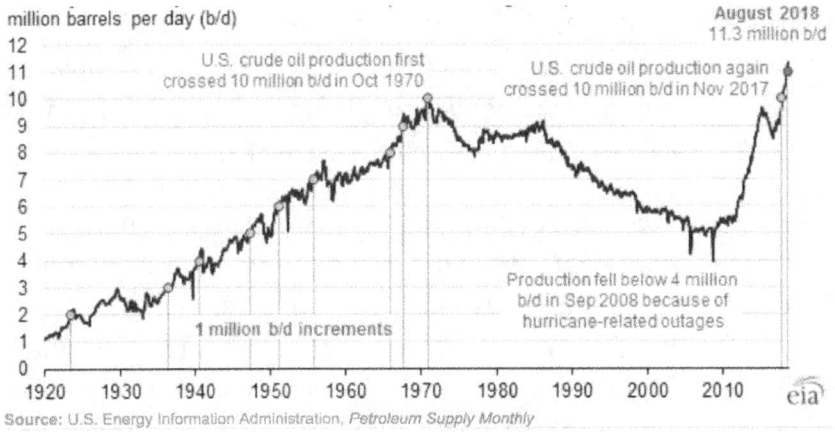

Figure 71 Monthly U.S. field production of crude oil

Remember, the experts said that American oil had peaked in 1970, and it appeared for a while that it had. That was until American inventors came along with the concept of fracking, and everything changed.

Saving the Global Environment

Is there something to be saved? Those of us in the developed world complain about the quality of our environment. Still, many of us remember the 1970s when America reached the peak of its environmental disaster. In the 1970s, Los Angeles was called by its more common name, "Smell-A," when the red alert days were averaging more than 200 days a year. During the same time, the Cuyahoga River, as it passed through Cleveland, caught fire at least a dozen times.[338] Things got so bad with the reckless disposal of litter that it led to a national TV campaign whose face was a crying Native American, played by an Italian-American actor.

Those days may be over in America, but they are not over in the developing world. In the 2020s, it is smog in cities like Cairo, Nairobi, and Kampala. While the media has been pushing the agenda with images of waste mounds from places like India,

Malaysia, and Myanmar. Trash pickers live from Shanghai to New York, with landfills slums existing in places like India, Ethiopia, and Brazil, to name a few.[339] People living in waste dumps is unheard of in the developed world but common outside of it.

It has been the United Nations and NGO's policy to provide the developing world with food and medicine to treat the symptoms of poverty rather than the causes of it. In the old adage, "Give a man a fish, and he eats for a day, but teach a man to fish, and he will never go hungry," the developed countries have preferred to give the developing world a fish rather than teach them to fish. At Southern Methodist University (SMU), research projects under a program of Engineers without Borders have been developing clean water methods for people in the Third World. Throughout the 20-year long project, the group has given clean water facilities to several small villages like the 150-person village of Llojila Grande[340] in 2018. They brought the village a fish. The fact is that they use the desperate needs of small villages for the purposes of research, more than resolving their needs.

While SMU's project was commendable, there is another way to bring clean water, in fact, clean water to the whole country. Safe, reliable, and cheap energy! The developed world gets clean water by water processing plants run by the grid and using the latest fossil fuel-derived materials. It is believed that 780 million people do not have access to improved water sources, and 1.8 billion do not have access to clean water.[341] At the same time, 940 million do not have access to electricity at all, and a further 2 billion to reliable electricity[342]. Are these numbers just coincidental? Actually, they are largely the same people, those that are impoverished, lack electricity and food.

How to save the environment? The answer is simple, wealth. Experience has shown that it is not government agencies like the EPA that have created clean environments, as Americans have both clean and dirty environments, all under the same EPA regulation structure. But what is different between those areas is wealth. Organizations like Cities4Forest[343] have the right ideas connecting the city's wealth with the tree canopy and green space. However, their premise is that give a town a green space with trees, and it will become wealthy. But instead, they should be making a city wealthy, and the residents will put in those green spaces and trees.

In 1997, Professor Paul Ekins of the University of London published a paper entitled, *"The Kuznets Curve for the Environment and Economic Growth"* in the *Journal of Environment and Planning*, which performed a mathematical analysis of how a country's per capita wealth impacted its environmental performance. The longstanding argument, originating with Thomas Malthus' "An Essay on the Principle of Population" and subsequently revamped in the 1970s by the Club of Rome's **LIMIT TO GROWTH,** was that as a nation got more affluent, the environment would get worse. It was assumed that wealth generated from industrialization could do little more than pollute. From the beginning of the industrial age until the 1970s, this seemed to be the case. The United States put in place the Environmental Protection Agency and lots of other regulations to get pollution under control, and it seemed to have worked. Still, Professor Ekins' work showed that there is a point that once a society reaches a certain income (seemingly near $10,000 per capita in 2020 dollars), the environment will improve naturally as the residents will have their basic needs met and will demand and work for improvement in their local environment. Thus, there becomes an inverted U-shaped curve that describes the relationship between wealth and the environment.

Therefore, the key to a better environment is to push as much of the world into the top half of the curve, which will be far more effective than rules and regulations. Deny these nations wealth, and they will never ascend to environmental improvements.

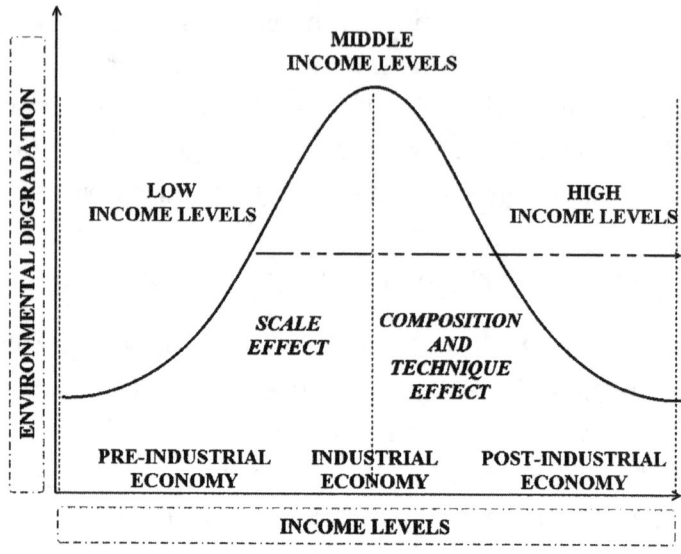

Figure 72 Kuznets Curve for the Environment[344]

The technological way out

The question is, "Who feeds Paris?" In 2011, *Regional Environmental Change Journal* published an article, "Grain, meat and vegetables to feed Paris: where did and do they come from?" which provides an analysis of the historical history of food consumption in the French capital.[345] While it is possible to track which cropland in northern France produces the grain and how the beef is produced mainly in Brazil, it fails to ask the critical question, who organizes all that food production for the 2.1 million residents? The answer, of course, is no one. Each market, restaurant, household, and bakery makes its own plans. Each orders the materials that fit their needs, at prices that are appealing, and each generally receives what it wants. It is the free market, Adam Smith's invisible hand, that feeds Paris. Too little of something, and consumers either pay more or choose something else. Too much of other things and daily specials arise featuring the abundant items.

So the question is, "who is responsible for America's energy and emissions policy?" As energy policy affects every aspect of life, stretches into nearly every industry, and represents billions of individual points of emission, there can be no one person. However, when no one person is responsible, it tends to be that no one takes responsibility. Psychologists call this *diffusion of responsibility*. A phenomenon first identified with the attack and rape of Kitty Genovese in a Queens, New York apartment building in which though many witnessed the attack, no one came to her aid or phoned for help. We now see this phenomenon frequently as it has become almost standard behavior for bystanders to record events on a cell phone rather than providing aid, which is the source of much of our nightly news footage. Energy policy influencers range from government regulators, industry executives, and technology developers down to the individual consumer of products and power. Each of these influencers are affected by the diffusion of responsibility phenomenon as each acts independently in their own best interest without regard to the larger implications of their actions. The government regulator may target international treaty compliance, the industry executive attempts to maximize market advantages, the developer wants to sell his technology and the consumer wants to relax and enjoy his weekend. Each player pushes certain behaviors optimized for themselves, yet the world benefits when those behaviors play out for the benefit of the national and international goals. It often isn't clear whether that diffusion of responsibility is a benefit or a defect when it comes to energy policy.

We have been living in a world where the diffusion of responsibility has been playing out in all sorts of ways. As Evan Sayet points out, "The Globalist / Socialist whom all flew off in private jets, to stay at five-star-hotels, typically on the monies taken from hardworking people, put together a massive one-size-fits-all global scheme to reduce carbon emissions. America was the only nation in the entire world not to sign [the Paris Accords]. America was also the only nation in the entire world to meet and exceed the reductions the scheme called for. Had Americans been forced to comply with the restrictions and the regulations conjured by the out of touch and wholly immune elite, Americans would not have been free to innovate, create, and make smarter local decisions based on local realities."[346]

The pollution of the 1970s has largely been eliminated without a massive upheaval in the economy. In the past 40 years, the population of America grew 45 percent while the GDP increased 5-fold, all while improving air quality: lead down 99%, carbon monoxide down 85%, sulfur dioxide down 92%, nitrogen oxides down 62%, industrial haze down 46% (PM10: larger particulate) and 43% (PM2.5: smaller particulate) and ozone down 35%.[347]

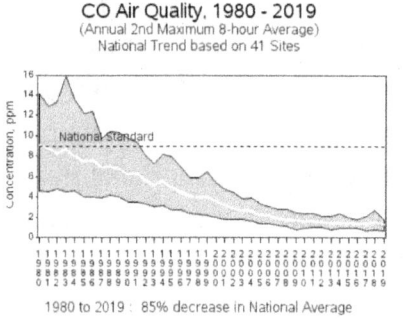 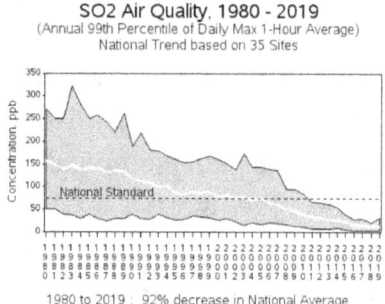

Figure 73 CO Air Quality **Figure 74 SO2 Air Quality**

The *Emissions Gap Report 2020*, the UN's annual assessment of greenhouse gas emissions noted that the United States with 13% of emissions was the "most successful major country at mitigating its own pollution" and "greenhouse gas emissions per capita in the U.S. are dropping precipitously while those of China, India and Russia continue to rise." Within the U.S. the GHG emissions have been declining by 0.4 percent per year.

Every proposed plan providing a path forward with the reduction of fossil fuels requires a significant drop in the metabolic rate for the nation. All those players in the energy game have, without coordination, influenced change there as well. Like those pollution levels that have fallen rapidly, the metabolic rate has been dropping as well. In 1975, it took 12,740 BTUs of energy to create a dollar of GDP in the United States. By 2019, that dropped to just 5,250 BTUs of energy. That is a drop of 60% in the metabolic rate. So, is it possible to have it fall another 50% from our present level? With the right technological path, it should be. Hopefully, that drop won't require eliminating the reliable, plentiful, and cheap electrical grid.

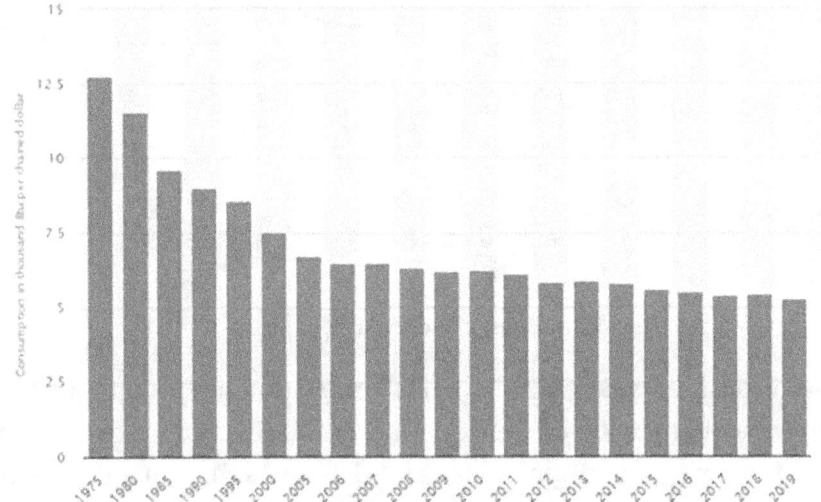

Figure 75 Energy Consumption per Real Dollar of GDP in the U.S.

So, the next question is, "Who plans America's Innovation?" Basically, the same person who plans the food for Paris and America's energy policy, no one. What feeds innovation? It has been driven by consumer demand, competition, and the relentless pursuit of advancement, achievement, and profits. Our innovation culture is one that accepted high risk – high reward, fierce competition, and ambitious drive for success.

Did America have an "Innovation Manifest Destiny?" Is it assured that the future will bring yet more innovation? The past may be revealing. Americans responded to the threat of fascism with such military-industrial output that they equipped not just themselves but many other nations. They responded to Sputnik by the pursuit of technical achievements that put a man on the moon. Fighting the Cold War with the Soviet Union demanded increased computing power, so America became the leader in computer platforms from mainframes to personal computers to microcontrollers to support everything from communication to weaponry.

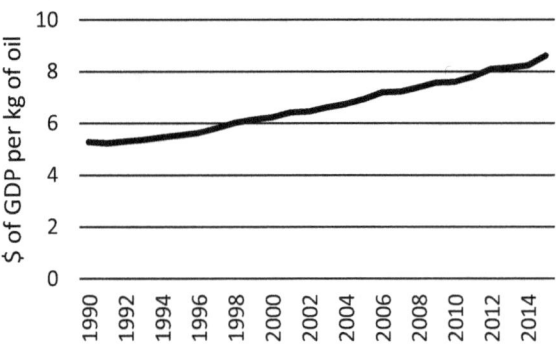

Figure 76 GDP per Unit of Energy Use - United States

In the past few years, technological innovation has brought us help on the demand side of the equation in the form of more efficient electric cars, smart appliances and devices, and demand leveling strategies. While at the same time, there were innovations on the energy generation side, including improved wind turbines, lower-cost solar and storage systems, along with the development of fracking and advanced oil and gas extraction.

Who plans innovation? The free market does. If the market is left free to boost worthy ideas while allowing unworthy concepts to end, America may well see more innovation pop up than ever before. When government entities attempt to plan innovations, we end up with projects like the solar panel company Solyndra and electric car manufacturer Fisker. Each was chosen as the companies of the future only to fail and enter bankruptcy. Government dollars and government bailouts did nothing to fix situations where the innovations did not meet the market need. Governments worldwide are being tempted to interfere due to the urgency of the climate crisis, but the innovation they crave will only come into being if they leave it alone. When the U.S. Government chose wind power as the winner and provided subsidies, wind power builds grew and when the subsidies went away so did the building.[348]

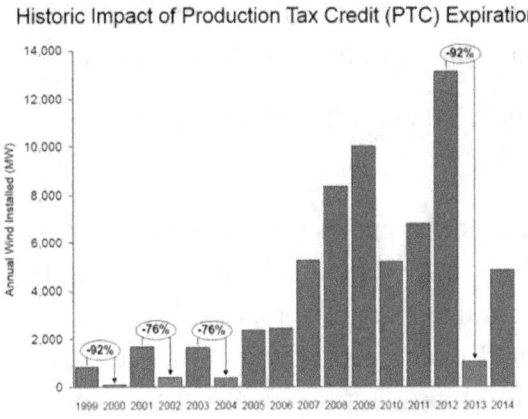

Figure 77 Wind Power Building Goes Away without Subsidies

The future of energy will depend on the innovations that will come our way. What does the Valentine's Day Winter Storm of 2021 teach us about how innovation in our electricity sources should develop in the future? Well, we all learned that wind and solar will never be able to provide power during the worst of winter storms. While wind and solar can play a role in delivering power into the electrical grid, it is unrealistic and even hazardous to rely upon them.

While wind and solar have a role to play in supplementing thermal energy production by saving fuel when the wind blows, the sun shines, and the weather is moderate, taking thermal production plants offline in favor of renewable sources is simply short-sighted and treacherous. Until we have a new technology to replace natural gas and nuclear plants, we will need to rely upon them for winter heat and summer cooling.

It would be unwise to try to guess and pick the winners ahead of time, rather the government must aid in making it profitable for innovators to come up and try out new ideas, either through subsidies or prize money for such discoveries.

Is there hope?

> *"Will prophecy come true? Perhaps not, but wagering against human ingenuity has always been a bad bet."*[349]
>
> Ronald Bailey, **THE END OF DOOM** (2015)

The struggle for answers comes down to two choices: dynamism or catastrophism[350]; a choice between things will get better because man will work through the problem, or things will worsen because there is no hope. History has shown that there is always hope, no matter how bad things appear, man has always found a way through. As Jeff Goldblum's character Dr. Ian Malcolm notes, in the movie *Jurassic Park*, "Life finds a way," as he observes that life always seems to live on no matter what disaster occurs.

From a catastrophism point of view, economist Martin Weitzman of Harvard University has put forward a theory, called the dismal theorem, that some things are so bad, so catastrophic, so dangerous to humanity that no amount of effort or money should be spared to pursue the fix.[351] In a sense, the pandemic response to the Coronavirus in 2020 has been a prime example of this theory. Lockdown the countries, take the hit to the economy, accept the 30% unemployment, stop educating the children. Basically, stop the world on its axis for a year or two, if necessary, to save us all.

But as we have learned through the lockdowns of 2020, this has consequences, and as William Nordhaus comments, "So if we accept the Dismal Theorem, we would probably dissolve in a sea of anxiety at the prospect of the infinity of infinitely bad outcomes."[352] Nordhaus's position is that climate change is just one of many possibilities of calamity that could face humanity and that "there is very little that we can rule out with 100 percent probability in most areas."[353]

Alternatively, with dynamism, Gregg Easterbrook points out, "it hardly promises that we'll approve of the future – only that we'll be able to live in it, and that a better world is coming."[354] The arrow of history always seems to have pointed to improvements. Through the darkest times of man's history, the fall of the Roman Empire, the black death, the world wars, communism, the nuclear age, in all cases, humanity has come out of the disasters in better shape than he went in. That is not to say that the millions of tragic deaths did not occur. But what rose from the ashes was more freedom, better living conditions, and longer, better lives. The alarm over the hole in the ozone over Antarctica is a perfect example of dynamism's success. First reported by Joseph Farman in 1985, the world had solved the problem by eliminating CFCs and implemented a treaty by 1987. The important thing to note is that engineers and scientists went to work immediately and found new compounds that could be used, in the same way with the same equipment, so that the troublesome CFCs could be eliminated. The solution was the result of engineering advances. The engineers solved the problem before the world governments could consult and come up with a governmental solution. In the end, governments merely implemented that which the engineers had already solved.

The way forward from the crossroads in which we find ourselves is to adopt a national strategy for energy, both production, and consumption. There are many paths of advancement, and there is not going to be one single solution, and all alternatives have advantages and disadvantages. Too often, so-called national strategies are a wish list of dreams and goals. Although the comments are directed at businesses, the remarks made by Freek Vermeulen, Professor of strategy and entrepreneurship at London Business School, apply to the energy strategies as well, "One major reason for the lack of action is that "new strategies" are often not strategies at all. A real strategy involves a clear set of choices that define what the firm is going to do and what it's not going to do."[355]

Responses to catastrophism can lead to proposals that could lead the world off the cliff in the name of disaster. Still, as long as there are those who will warn us, we may be saved. Engineers of the past have alerted us to potential disasters such as levee failures in New Orleans, road collapses in Oakland, and defective rivets on the Titanic. Each of those warnings were not heeded but engineers continue to raise alarms. As long as engineers like Allan McDonald raise the warnings, there will be hope. While his bosses didn't listen, Mr. McDonald, a Morton Thiokol engineer, warned against launching the space shuttle Challenger on that cold January morning of 1986 due to risks previously seen on the O-rings. His stance was "the smartest decision I ever made in my lifetime." His warning could have saved those seven lives. Even still, his objections assured that the issue was repaired and future shuttles were no longer at risk, at least not for that particular defect. In America today, there are many more Allan McDonalds warning of issues with the emerging technologies upon which we are basing our new energy transition. They warn of the resource limitations, the impracticality of renewables, and the loss of those wonder chemical products. Will we listen to their warnings and let them develop the right way forward? If the world listens, the alarms can steer us all in the right direction.

Whether you believe that American is an exceptional nation or not, the one thing we must all admit is that America has had a history of exceptional engineering, whether the transcontinental railroad, Panama Canal, the age of invention, the space race, or high technology. American engineers have always led the way to the future, and our current future will be no different. Imagine back 30 years and all the gadgetry and technologies that we rely upon today which didn't exist then: GPS, robotic surgery, smartphones, smartwatches, smart pacemakers, DNA testing, social media, online shopping, and the solar and wind power industry. What will the next 30 years bring?

As George Washington wrote to Charles Thruston in 1774, "Truth will ultimately prevail where pain is taken to bring it to light." The time may have indeed come when the uphill struggle associated with bringing the truth to light may be worth it. While we face the risk of censorship, cancellation, and rejection in speaking up, the fact is that what we have is worth saving. Against the platitudes of the media and environmental activists, there is a positive story to tell, which too often goes untold. During the past 170 years, America's innovation has taken a nation from an isolated agrarian society of little impact to the world's superpower. Becoming the country that provides more aid and assistance to other nations than any other. The nation that exported democracy to the world. It is the United States that shared with the rest of the world all those wonderful improvements to life that are worth saving. The developed nations have risen to the top of the heap due to those wonder petrochemicals, the positive benefits of the reliable, plentiful, and cheap electrical grid, and the ever-present energy sources which have led to longer, healthier, safer, and more productive lives. The rest of the world will benefit more from sharing those wonders, rather than denying everyone those benefits in the name of the environment. And finally, as Ronald Reagan said best, "You and I have a rendezvous with destiny. We will preserve for our children this, the last best hope of man on earth, or we will sentence them to take the first step into a thousand years of darkness."

13

A Proposal

I began this project with an open mind about the proper course forward, which technologies should be pursued, and what strategies would be the right ones. Throughout my extensive reading and research for this book, I have developed opinions, although I would never say that they are more than opinions or that I have all the answers. This chapter will outline some of the better ideas for moving forward to a brighter future for humanity, as I see them.

The best news of all on this topic is that basically, most proponents of the war on fossil fuel aren't truly serious about the end goal. The best examples I can point to is that on Earth Day 2021, President Biden spoke at the virtual climate summit indicating that independent of the actions of other countries, America would cut her emissions in half by 2030, and later he spoke at a news conference about his infrastructure proposal to rebuild roads and bridges over those same years, even when construction is the largest single emission source. Further, while pushing electric vehicles, the Biden Administration has also placed limits on mining of lithium and other critical materials needed for those vehicles. The war on fossil fuels is more about politics than it is about action. Witness the fact that Bill Gates spends $7 million annually to offset his carbon footprint, rather than sacrifice to reduce it. Even while selling his book, **HOW TO AVOID A CLIMATE DISASTER**, he has been bidding to purchase Signature Aviation, which provides ground services for 1.6 million private jet flights a year.

My proposal for how to deal with the climate and the need to wean ourselves off of fossil fuels comes under two themes:
1) Teach Nations to Fish
2) Let Technology Win

Teach Nations to Fish

As we have seen with the Kuznets Curve in Chapter 12, the best way to get the developing and undeveloped nations to be concerned with environmental issues is to increase their per capita wealth. While the United Nations has a history of pursuing ecological concerns in the name of the lesser countries, this is generally for the purpose of handouts. Rather than providing these nations "fishes," it is time to teach these nations to fish. The primary issue with feeding the nations is that technology suites are not developed to sustain the nation. Those post-colonial nations still have difficulties releasing the grip on their former colonies, especially in Africa, as evidence by France's control over the monetary systems of the West African Nations. The UN setting up camps to distribute supplies is not nearly as useful as it would be for the country itself to do the work of importing and distributing supplies as this helps to establish suites.

The IPCC intends that those 2.5 billion people who burn wood and animal waste for heating and cooking should leapfrog the fossil fuel stage in favor of renewables. However, with the technology available today, moving these nations to fossil fuels, even coal, would reduce their emissions relative to wood burning. The introduction of uncontrollable, unreliable electricity does not discourage traditional methods as they need them when electricity is not available. Many of those still using traditional methods have access to electricity but cannot depend on it due to its unreliability. Only reliable, plentiful, and cheap electricity can do that, and the only practical way to do it is for most places, coal and, for a few, natural gas. As opposed to the current plan to move to renewables, the developed nations should begin initiatives that encourage undeveloped countries to build fossil fuel infrastructures, including reliable power generation grids. As Shellenberger pointed out, burning coal is better than burning wood, and burning natural gas is better than burning coal.

By providing fossil fuel technologies, assistance with building coal-powered electrical plants, and power line infrastructure, the developed nations can help less developed countries get to that magical point where the washing machine becomes a viable choice for the home, and the lives of all are changed forever.

Let Technology Win

While public-private partnership arrangements are the latest trend in directing technology development by the government, its track record is mixed, succeeding in some places like the Covid Vaccine while failing in other places like Solyndra and numerous correction facilities. Although, in some camps even the vaccine should be counted as a loser. The government often has a bad track record of picking the right winners when picking the winners and losers.

In the past, international prize competitions have pushed the envelope of technologies in all sorts of fields. Some of the achievements due to these prizes include:

- The Marine Clock
- Canned Foods and Margarine
- The Steam Tractor
- The Steam Locomotive
- Polar Exploration
- Transatlantic Flight
- Human Powered Airplanes and Helicopters
- The solution to Fermat's Last Theorem

- Computer Chess
- Artificial Intelligence
- Private Space Flight

The most efficient way to get the needed technology without weighing the scales towards a particular technology would be to support a series of international prizes.

Some prize ideas are:

- New Battery Technologies that use less raw materials.
- Improvements in solar panel constructions
- Direct Current Powered Solutions
- Superconductors for Power Transmission
- Developments in alternative fuels for transportation, such as hydrogen and LPG.
- Home Heating Technologies

As Thomas Edison once said, "When you have exhausted all possibilities, remember this, you haven't." Maybe it's time to look for more possibilities.

So that the right technology can win, it is essential to level the playing field so that nuclear encounters a fair field. The regulations around nuclear power are designed to make the idea of installing a facility essentially illegal. Nuclear is the only non-carbon independent, controllable, and reliable power generation source currently available. There have been three significant accidents in the past, however, each occurred with out-of-date, older technologies which would not be in use today. Further, the two accidents not caused by earthquake and tsunami were caused by human error during experimentation. Newer generation nuclear reactors are smaller, safer, and cheaper. The cost of America's nuclear facilities is inflated due to their essentially one-off design process. Engineers from South Korea have developed a method of producing reactors while not exactly mass production, it does gain the cost benefit of scale. Using their experience and knowledge would reduce the cost of future American reactors considerably. The closing of Vermont Yankee, Indian Point, and Diablo Canyon nuclear facilities only has the effect of increasing net emissions.

The rate at which new technologies are adopted is controlled by the cost to integrate the new devices into our existing infrastructure. To reduce the cost of adopting future ideas into homes, building codes should be modified so that homes will be more change-ready, including installing smart electrical panels that can accept alternative power sources, automatically switch to backup power, and ready for EV charging units. Such smart homes would be equipped with devices that control electricity usage based on ongoing grid demand. Appliances such as washing machines, dryers, and dishwashers would check the grid status via the internet and postpone the operation until sufficient margin exists on the grid.

Government direct payments and tax incentives disrupt the ability of the market to determine the winning technologies. To date, the policymakers have deemed solar and wind as those technologies that must win and, as a result of their incentives, have accelerated their installation beyond the ability of the grid to manage them, the same is true for electric vehicles. The capitalism marketplace represents millions of small decisions by experts, technicians, and engineers, each making the best decisions on a step-by-step basis. When left to do its work, the market will find the best winners. And who knows, when the technologies suites are ready, wind and solar may win after all, or maybe it will be a as yet unknown technology.

Epilogue

The major obstacle to writing about any topic which is an ongoing concern, is that there is always more information to share. Everyday there are new stories, new ideas being floated and new proposals, each worth discussion but time comes when the book must be completed. Mostly, the book's contents are safe as while new ways of expressing the concepts are being created, the ideas and proposals aren't changing significantly and are unlikely to change in the near future.

The world seems obsessed with going off the cliff in favor of uncontrollable, unreliable, intermittent energy despite all the warning signs and all the evidence to why this is a bad idea. One of the lessons of the pandemic of 2020 is that politicians may still go places in the name of science when science actually says, don't go there. Reading about the pandemic panics of the past reveals that the recent actions of politicians reflect the ideas of the past, rather than allowing the marketplace to find new solutions.

Within their range of influence, the readers should do what they can to educate those around them, but, at a minimum, the reader should prepare themselves to not go off the cliff with the rest of the world. The future will most likely be a time of increasingly intermittent electricity, especially during times of peak loading when it is needed the most. Every household should be ready with backup power in whatever form is appropriate: battery, solar, fuel generator, or wind, along with alternative sources of life-saving heat: propane or space heating for example. Whatever your plan, have a plan. As Benjamin Franklin printed in his Farmer's Almanac, "failure to plan is planning to fail." With knowledge and planning, you may be lucky enough to not fall victim to the coming dark age.

So what is the likelihood that the world goes down the path of banning fossil fuels? The influential NGO, the International Energy Agency, has released widely distributed reports that demands that all oil and gas production be shut down immediately or there will be no other way to avoid climate change. On May 25, 2021, a Dutch Court used that report when it ruled that Shell "must cut its greenhouse gas emissions more aggressively; by 2030, Shell's net carbon emissions needed to be 45% lower than 2019 levels." [356] Judge Alwin ruled that "the interest served with the reduction obligation (climate change) outweighs the Shell group's commercial interests" while acknowledging that "a change of policy from Shell could curb the potential growth of the Shell group." Angus Walker, a London based environmental lawyer said, "This may spread from large emitters to small, and from the Netherlands to other countries, at least in terms of challenges, if not successful ones." Who will push the world off the cliff? It could be politicians, non-profits, corporations, or the court system, each influenced by their own special interests.

As a December 23, 2020 op-ed in the Daily Sentinel put it, "Let's face it: we've all got oil on our hands. Fossil fuels are intertwined with every modern convenience known to man. Fertilizers, medicine, appliances, food packaging — the list goes on and on — reflect a massive dependence on hydrocarbons. We have written [op-eds] many times about the enormous benefits to mankind — particularly in Third World food production — afforded us by the oil and gas industry. We can't simultaneously enjoy these things and condemn the industry that delivers them to us."

Some will continue to push the world towards the next energy transition. Still, the world we observe around us all shows that the conditions of life are improving everywhere for everyone. If allowed to continue, the future will be bright.

Endnotes

[1] Toffler, Alvin, FUTURE SHOCK (New York, Random House, 1970) pg. 15
[2] The Atlantic, "The 50 Greatest Breakthroughs since the Wheel", Nov. 2013, www.theatlantic.com/magazine/archive/2013/11/innovations-list/309536/
[3] Stein, Ronald; Royal, Todd. JUST GREEN ELECTRICITY: HELPING CITIZENS UNDERSTAND A WORLD WITHOUT FOSSIL FUELS (p. 48). Archway Publishing. Kindle Edition.
[4] American Petroleum Institute, "Energy and the 2020 Election", www.api.org/news-policy-and-issues/american-energy/energy-and-the-2020-election
[5] Stein & Royal, Page 88
[6] Mohammed Abbas & Repudaman Singh. (2014). *A Survey of Environmental Awareness, Attitude and Participation Amongst University Students: A Case Study*. International Journal of Science and Research (IJSR). 3. 1755-1760.
[7] Yeh, Shin-Cheng; Huang, Jing-Yuan; Yu, Hui-Ching. 2017. "*Analysis of Energy Literacy and Misconceptions of Junior High Students in Taiwan*" Sustainability 9, no. 3: 423. doi.org/10.3390/su9030423
[8] http://priceofoil.org/2021/03/18/letter-biden-public-finance-fossil-fuels/
[9] Galef, Julia. THE SCOUT MINDSET (pp. 178). Penguin Publishing Group. Kindle Edition.
[10] Stein, R., & Royal, T, Page 53
[11] Greer, John Michael. DARK AGE AMERICA: CLIMATE CHANGE, CULTURAL COLLAPSE, AND THE HARD FUTURE AHEAD. New Society Publishers, 2016. (pp. 4-5)
[12] Discover, "Just How Dark Were the Dark Ages?", www.discovermagazine.com/the-sciences/just-how-dark-were-the-dark-ages
[13] Tamny, John. WHEN POLITICIANS PANICKED: THE NEW CORONAVIRUS, EXPERT OPINION, AND A TRAGIC LAPSE OF REASON (p. 165). Post Hill Press. Kindle Edition.
[14] Idib p. 165

15 The Department of Energy, "2017 U.S. Energy and Employment Report", www.energy.gov/downloads/2017-us-energy-and-employment-report
16 American Psychological Association. (n.d.). APA Dictionary of Psychology. dictionary.apa.org/confirmation-bias
17 Hoffman, Bryce G. **RED TEAMING: HOW YOUR BUSINESS CAN CONQUER THE COMPETITION BY CHALLENGING EVERYTHING.** Crown Business, 2017.
18 Guinness, Os. **THE DUST OF DEATH.** Intervarsity Press, 1975. Page 309-310
19 Noor, Iqra, "Confirmation Bias", *Simply Psychology*, www.simplypsychology.org/confirmation-bias.html
20 Cherry, Kendra, "How Confirmation Bias Works", *Very Well Mind*, www.verywellmind.com/what-is-a-confirmation-bias-2795024
21 Kahan, D., Peters, E., Wittlin, M. et al. "The polarizing impact of science literacy and numeracy on perceived climate change risks." *Nature Clim Change 2*, 732–735 (2012). doi.org/10.1038/nclimate1547
22 NASA, "Scientific Consensus: Earth's Climate is Warming", climate.nasa.gov/scientific-consensus/
23 Galef, Julia. Page 5-6
24 http://jmhartley.com/snell/snellupd0001.htm#id2193
25 Cummins, Neil, "*Longevity and the Rise of the West*", London School of Economics and Political Science, 2014
26 www.macmillandictionary.com/us/dictionary/american/decarbonize
27 Mahbubani, Kishore. *The Great Convergence: Asia, the West, and the Logic of One World.* Publicaffairs, 2014.
28 Schneider, Stephen, Editorial, 22 Nov, 1989, *Detroit News*, stephenschneider.stanford.edu/Publications/PDF_Papers/DetroitNews.pdf
29 James D. Agresti, Rachel McCutcheon, Steven Bukovec and Schuyler Dugle, "Global Warming Facts," *Just Facts*, last modified December 20, 2020, www.justfacts.com/globalwarming.
30 Agresti, James D, "Scientific Survey Shows Voters Widely Accept Misinformation Spread by Media", *Just Facts*, www.justfacts.com/news_2019_survey_voter_knowledge
31 Tax Foundation, "Summary of the Latest Federal Income Tax Data, 2020 Update", taxfoundation.org/summary-of-the-latest-federal-income-tax-data-2020-update/
32 Greer, John Michael. **DARK AGE AMERICA: CLIMATE CHANGE, CULTURAL COLLAPSE, AND THE HARD FUTURE AHEAD.** New Society Publishers, 2016.
33 Kathleen Ronayne and Ellen Knickmeyer, "Biden calls for 'transition' from oil, GOP sees opening", *Associated Press*,

apnews.com/article/election-2020-joe-biden-donald-trump-technology-climate-26908b855045d5ce7342fd01be8bcc10
34 The Solutions Project, thesolutionsproject.org/
35 Goldman School of Public Policy, University of California Berkley, *The 2035 Report*, www.2035report.com/electricity/
36 Sanders, Bernie, berniesanders.com/issues/green-new-deal/
37 Biden, Joe, joebiden.com/wp-content/uploads/2020/08/UNITY-TASK-FORCE-RECOMMENDATIONS.pdf
38 Donaghy, Tim, "4 ways to shrink the fossil fuel industry and protect the climate", *Greenpeace*, www.greenpeace.org/usa/4-ways-to-shrink-the-fossil-fuel-industry-and-protect-the-climate/
39 United Nations Climate Change, IPCC, unfccc.int/cop25
40 U.S. Energy Information Administration, "Electricity Explained" www.eia.gov/energyexplained/electricity/electricity-in-the-us.php
41 Comptroller, State of Texas, "Texas Electricity Resources", comptroller.texas.gov/economy/fiscal-notes/2020/august/ercot.php
42 California Energy Commission, www.energy.ca.gov/data-reports/energy-almanac/california-electricity-data/2019-total-system-electric-generation/2018
43 United States Geographical Survey, U.S. Department of the Interior, "How much wind energy does it take to power an average home?" www.usgs.gov/faqs/how-much-wind-energy-does-it-take-power-average-home?qt-news_science_products=0#qt-news_science_products
44 USGS, The U.S. Wind Turbine Database, eerscmap.usgs.gov/uswtdb/
45 U.S. Department of Energy, "U.S. Installed and Potential Wind Power Capacity and Generation", windexchange.energy.gov/maps-data/321
46 Sanders, Bernie, berniesanders.com/issues/green-new-deal/
47 U.S. Energy Information Administration, "Nuclear Explained", www.eia.gov/energyexplained/nuclear/us-nuclear-industry.php
48 World Nuclear Association, "Plans for New Reactors Worldwide", www.world-nuclear.org/information-library/current-and-future-generation/plans-for-new-reactors-worldwide.aspx
49 Statista, "Projected electricity generation capacity worldwide from 2018 to 2050", www.statista.com/statistics/859178/projected-world-electricity-generation-capacity-by-energy-source/
50 Statista, "Distribution of electricity generation worldwide in 2018, by energy source", www.statista.com/statistics/269811/world-electricity-production-by-energy-source/
51 International Atomic Energy Agency, "Preliminary Nuclear Facts and Figures for 2019", www.iaea.org/newscenter/news/preliminary-nuclear-power-facts-and-figures-for-2019

[52] Storrow, Benjamin, "Coal's Decline Continues with 13 Plant Closures Announced in 2020", *Scientific American*, www.scientificamerican.com/article/coals-decline-continues-with-13-plant-closures-announced-in-2020/
[53] Sanders, Bernie, berniesanders.com/issues/green-new-deal/
[54] NaturalGas.Org, http://naturalgas.org/overview/uses-residential/
[55] National Conference of State Legislatures, www.ncsl.org/research/energy/state-gas-pipelines.aspx
[56] International Panel on Climate Change, www.ipcc.ch/2019/08/08/land-is-a-critical-resource_srccl/
[57] The Geography of Transport Systems, transportgeography.org/contents/chapter8/urban-land-use-transportation/
[58] Montanez, Amanda, "How much of the world's protected land is actually protected, *Scientific America*, www.scientificamerican.com/article/how-much-of-the-worlds-protected-land-is-actually-protected1/
[59] Carrington, Damian, "Avoiding meat and dairy is single biggest way to reduce the impact on earth", *The Guardian*, www.theguardian.com/environment/2018/may/31/avoiding-meat-and-dairy-is-single-biggest-way-to-reduce-your-impact-on-earth
[60] McMurry, Patty, "Joe Biden's Far-Left Climate Plan Seek to Limit Americans to 4 Lbs of Beef Per Year", 100percentfedup.com/joe-bidens-far-left-climate-plan-seeks-to-limit-americans-to-4-lbs-of-beef-per-year/
[61] https://markets.businessinsider.com/news/stocks/iea-drops-bombshell-report-on-oil-and-gas-1030443414
[62] Blackrock Investments, www.blackrock.com/au/individual/blackrock-client-letter
[63] Partnership for Carbon Accounting Financials, "The Global GHG Accounting and Reporting Standard for the Financial Industry", carbonaccountingfinancials.com/standard
[64] Smith, Abby, "Joe Biden considers ordering climate confessions", *Washington Examiner*, April 12, 2021, www.washingtonexaminer.com/policy/energy/biden-eyes-order-climate-disclosures
[65] Energy Information Administration, www.eia.gov/tools/faqs/faq.php?id=97&t=3
[66] Independent Petroleum Association of America, www.ipaa.org/independent-producers/

Endnotes

[67] ExxonMobil, corporate.exxonmobil.com/Sustainability/Emissions-and-climate

[68] Breakthrough Energy, www.breakthroughenergy.org/our-story/our-story

[69] Breakthrough Energy, www.breakthroughenergy.org/

[70] Institute for Applied Ecology, "How additional is the Clean Development Mechanism?", DG CLIMA, ec.europa.eu/clima/sites/clima/files/ets/docs/clean_dev_mechanism_en.pdf

[71] Dhillon, Amrit, "How Coldplay's green hopes died in the arid soil of India", *The Telegraph*, April 30, 2006, www.telegraph.co.uk/news/worldnews/asia/india/1517031/How-Coldplays-green-hopes-died-in-the-arid-soil-of-India.html

[72] Struck, Doug, "Carbon Offsets: How a Vatican forest failed to reduce global warming", *Christian Science Monitor*, www.csmonitor.com/Environment/2010/0420/Carbon-offsets-How-a-Vatican-forest-failed-to-reduce-global-warming

[73] Clifford, Catherine, "Bill Gates on Climate Change: I'm another 'rich guy with an opinion' – but here's why you should listen", *CNBC*, www.cnbc.com/2021/02/14/bill-gates-on-his-carbon-footprint.html

[74] Amazon, sustainability.aboutamazon.com/

[75] McLaren, Duncan, "The problem with net-zero emissions targets", *Carbon Brief*, www.carbonbrief.org/guest-post-the-problem-with-net-zero-emissions-targets

[76] "Betting on Negative Emissions", *Nature Climate Change*, Oct. 2014 pages 850-852

[77] University of Leeds. "Shining a light on international energy inequality." *ScienceDaily*, 16 March 2020. www.sciencedaily.com/releases/2020/03/200316141505.htm

[78] *Plano Star Courier*, March 21, 2021 page 1-3

[79] Bellis, Mary, "American Farm Machinery and Technology Changes from 1776-1990", *ThoughtCo.* , www.thoughtco.com/american-farm-tech-development-4083328

[80] Morris, Glenn, **ABANDONED OHIO**, ohioghosttowns.org/Vinton-county/

[81] Stein, R., & Royal, T, Page 53

[82] ourworldindata.org/wrong-about-the-world

[83] Bailey, R., **THE END OF DOOM: ENVIRONMENTAL RENEWAL IN THE TWENTY-FIRST CENTURY**. New York: Thomas Dunne Books. 2015, Page 67

[84] bigthink.com/scotty-hendricks/there-are-more-than-100-uncontacted-tribes-in-the-world-who-are-they

[85] Cleaveland, Nancy, "Twisting Hay", *Laura Ingalls Wilder A-Z*, http://www.pioneergirl.com/blog/archives/6406

[86] Kitajima, Yasuko, "All Agricultural Production during 1990", http://www-formal.stanford.edu/jmc/nature/node15.html

[87] Ramaswamy, NS, "Draught animals and welfare", 1994. www.oie.int/doc/ged/D8880.pdf

[88] en.wikipedia.org/wiki/Slavery_in_the_21st_century

[89] "Indoor air pollution in developing countries: a major environmental and public health challenge", *Bulletin of the World Health Organization*, 2000 78(9), page 1078

[90] Energy Education, University of Calgary, energyeducation.ca/encyclopedia/Primary_fuel

[91] Energy Education, University of Calgary, energyeducation.ca/encyclopedia/Secondary_fuel

[92] Fischer-Kowaksi, M. & Haberl, H, "Social Metabolism: A metric for biophysical growth and degrowth", Oct. 2015

[93] Haberl, et.al. 2011)

[94] Ecowatch, www.ecowatch.com/96-cities-that-are-quitting-fossil-fuels-and-moving-toward-100-renewab-1882098115.html#toggle-gdpr

[95] Roberts, David, "The key to tackling climate change: electrify everything", *Vox*, www.vox.com/2016/9/19/12938086/electrify-everything

[96] en.wikipedia.org/wiki/World_energy_consumption

[97] EIA, www.eia.gov/energyexplained/us-energy-facts/

[98] EIA, www.eia.gov/energyexplained/electricity/use-of-electricity.php

[99] Magill, Bobby, "The Fuel you use for heating depends on where you live", *Climate Central*, www.climatecentral.org/news/your-heating-fuel-depends-on-where-you-live-18084

[100] Kovaleski, Dave, "Electricity is the most used power source for cooking in the U.S.", *Daily Energy Insider*, dailyenergyinsider.com/news/16144-electricity-is-the-most-used-power-source-for-cooking-in-the-us/

[101] Sinclair Oil Company, www.sinclairoil.com/dino-history

[102] Goldsberry, Clair, "Sorry, folks, oil does not come from dinosaurs", *Plastics Today*, www.plasticstoday.com/materials/sorry-folks-oil-does-not-come-dinosaurs

[103] Stein, R., & Royal, T., Page 60

[104] American Chemical Society, www.acs.org/content/acs/en/education/whatischemistry/landmarks/pennsylvaniaoilindustry.html

[105] Kimray Inc., kimray.com/training/types-crude-oil-heavy-vs-light-sweet-vs-sour-and-tan-count

[106] Penn State College of Earth and Mineral Sciences, www.e-education.psu.edu/fsc432/node/5
[107] Penn State College of Earth and Mineral Sciences, www.e-education.psu.edu/fsc432/content/paraffins
[108] www.oklahomaminerals.com/understanding-crude-oil-classifications
[109] www.law.cornell.edu/wex/sine_qua_non
[110] www.breakthroughfuel.com/blog/crude-oil-barrel/
[111] www.aaa.com/autorepair/articles/where-does-gasoline-come-from
[112] Burke, J. (1998). **The pinball effect: How Renaissance water gardens made the carburetor possible and other journeys through knowledge.** London: Little, Brown. Page 109-110
[113] Daimler, media.daimler.com/marsMediaSite/en/instance/ko/Wilhelm-Maybach-1846---1929.xhtml?oid=9904622
[114] Stein, R., & Royal, T., Page 60
[115] breakfree2016.org/
[116] www.greenpeace.org/usa/were-ready-to-break-free-from-fossil-fuels-are-you/
[117] en.wikipedia.org/wiki/Velocipede
[118] www.insider.com/birkenstock-eva-sandals-review
[119] American Fuel and Petrochemical Manufacturers, www.afpm.org/industries/products/petrochemicals
[120] Burke, James, CIRCLES, page 75
[121] Burke, James, THE PINBALL EFFECT, page 48
[122] cen.acs.org/articles/95/i12/Chinas-key-feedstock-above.html
[123] www.theguardian.com/environment/2021/feb/13/bill-gates-on-the-climate-crisis-i-cant-deny-being-a-rich-guy-with-an-opinion
[124] Owen, Josh, "Is it possible to make steel without fossil fuels?", *GreenBiz*, www.greenbiz.com/article/it-possible-make-steel-without-fossil-fuels
[125] www.bbc.com/news/science-environment-46455844
[126] Environmental Protection Agency, www.epa.gov/ingredients-used-pesticide-products/overview-wood-preservative-chemicals
[127] journalnow.com/lifestyles/food/silent-dangers-traditional-paints-solvents-emit-harmful-toxic-chemicals/article_4648cea8-8784-11e2-933c-001a4bcf6878.html
[128] Jeremy Hess, Et al., "Petroleum and Health Care: Evaluating and Managing Health Care's Vulnerability to Petroleum Supply Shifts", *American Journal of Public Health 101*, no. 9 (September 1, 2011): pp. 1568-1579. https://doi.org/10.2105/AJPH.2011.300233
[129] U.S. Tire Manufacturers, www.ustires.org/whats-tire-0

[130] Morris, Hugh, "Airline weight reduction to save fuel: The crazy ways airlines save weight on planes", *Traveller*, www.traveller.com.au/airline-weight-reduction-to-save-fuel-the-crazy-ways-airlines-save-weight-on-planes-h14vlh

[131] Zhang, Benjamin, "Another window on a Southwest plane has failed, but airplane windows are stronger than you think", *Insider*, www.businessinsider.com/airplane-windows-durable-safe-2018-4

[132] Marsh, Georgy, "Can trains be half plastic?", *Reinforced Plastics Materials Today*, www.materialstoday.com/composite-processing/features/can-trains-be-half-plastic-part-1/

[133] Williams ED, Ayres RU, Heller M. The 1.7 kilogram microchip: energy and material use in the production of semiconductor devices. *Environ Sci Technol*. 2002 Dec 15;36(24):5504-10. doi: 10.1021/es025643o. PMID: 12521182.

[134] www.eesemi.com/moldcomp.htm

[135] Taylor, Kate, *Insider*, www.businessinsider.com/first-lady-melania-trump-vs-michelle-obama-fashion-cost-of-clothing-2017-8#campaigning-at-political-conventions-1

[136] www.fibre2fashion.com/industry-article/3085/retail-uses-of-cotton

[137] textileexchange.org/

[138] www.gminsights.com/industry-analysis/polyester-fiber-market

[139] Thomas, Dana, "The High Price of Fast Fashion", *Wall Street Journal*, www.wsj.com/articles/the-high-price-of-fast-fashion-11567096637

[140] Johnson, Emma, "The Real Cost of Your Shopping Habits", *Forbes*, www.forbes.com/sites/emmajohnson/2015/01/15/the-real-cost-of-your-shopping-habits/?sh=4c8b79041452

[141] www.sustainablefashion.earth/type/water/synthetic-fibres-used-in-72-clothing-items-can-sit-in-landfills-for-200-years/

[142] http://www.madehow.com/Volume-5/Spacesuit.html

[143] SpaceX, www.spacex.com/vehicles/falcon-9/

[144] Stein, R., & Royal, T., Page 64

[145] petroleumservicecompany.com/blog/history-evolution-medicine-2/

[146] www.ncbi.nlm.nih.gov/pmc/articles/PMC3154246/

[147] context.capp.ca/articles/2019/feature_petroleum-in-real-life_pills

[148] Ibid.

[149] orionmagazine.org/article/medicine-after-oil/

[150] Jeremy Hess, Et Al., 2011: Petroleum and Health Care: Evaluating and Managing Health Care's Vulnerability to Petroleum Supply Shifts *American Journal of Public Health 101*, 1568_1579, doi.org/10.2105/AJPH.2011.300233

[151] Ibid.
[152] Ehrlich, Paul R. **POPULATION BOMB**. Ballantine Books, 1968.
[153] Epstein, Alex, Center for Industrial Progress, industrialprogress.com/fossil-fuels-are-the-food-of-food/
[154] Mclintock, Kaitlyn, "The Average Cost of Beauty Maintenance could put you through Harvard", *Byrdie*, www.byrdie.com/average-cost-of-beauty-maintenance
[155] www.statista.com/statistics/243742/revenue-of-the-cosmetic-industry-in-the-us/
[156] allusesof.com/energy/75-uses-of-fossil-fuels-in-daily-life/
[157] www.nationalgeographic.org/encyclopedia/wetsuit/
[158] Roberts, William C., "Facts and Ideas from Anywhere", Baylor *University Medical Center Proceedings*, Vol. 23, 2010, www.tandfonline.com/doi/abs/10.1080/08998280.2010.11928617
[159] Sayet, E, **THE WOKE SUPREMACY**, 2020, page 148.
[160] Farrand, Phil. **THE NITPICKER'S GUIDE TO THE NEXT GENERATION TREKKERS**. Dell, 1993.
[161] science.howstuffworks.com/10-reasons-space-exploration-matters.htm
[162] www.nasa.gov/sites/default/files/80660main_ApolloFS.pdf
[163] Stein, R., & Royal, T., Page 46
[164] ourworldindata.org/child-mortality
[165] Institute of Medicine (U.S.) Committee on Palliative and End-of-Life Care for Children and Their Families; Field MJ, Behrman RE, editors. Washington (DC): National Academies Press (U.S.); 2003. www.ncbi.nlm.nih.gov/books/NBK220806/
[166] Colchero, Et. Al., "The emergence of longevous populations." *Proceedings of the National Academy of Sciences of the United States of America* [First published online: 21 November 2016]. DOI:10.1073/pnas.1612191113
[167] www.census.gov/library/stories/2017/10/aging-boomers-deaths.html
[168] ourworldindata.org/economic-growth
[169] BBC, "Extreme poverty set for first rise since 1998, World Bank warns", www.bbc.com/news/business-54448589
[170] Worstall, Tim, "By global standards there are no American poor; all in the us are middle class or better", *Forbes*, www.forbes.com/sites/timworstall/2014/08/27/by-global-standards-there-are-no-american-poor-all-in-the-us-are-middle-class-or-better/?sh=5fdd650b5cb5
[171] www.actionagainsthunger.org/world-hunger-facts-statistics

[172] Erdman, Jeremy, "We produce enough food to feed 10 billion people. So why does hunger still exist?" medium.com/@jeremyerdman/we-produce-enough-food-to-feed-10-billion-people-so-why-does-hunger-still-exist-8086d2657539

[173] Associated Press, www.theguardian.com/environment/2011/nov/28/un-farmers-produce-food-population

[174] Indur M. Goklany, "Wealth and Safety: The Amazing Decline in Deaths from Extreme Weather in an Era of Global Warming, 1900—2010," *Reason Foundation Policy Study* No. 393, September 2011 reason.org/wp-content/uploads/files/deaths_from_extreme_weather_1900_2010.pdf

[175] IPCC, www.ipcc.ch/report/ar5/wg2/

[176] Tamny, John. **WHEN POLITICIANS PANICKED: THE NEW CORONAVIRUS, EXPERT OPINION, AND A TRAGIC LAPSE OF REASON** (p. 165). Post Hill Press. Kindle Edition.

[177] Green Age, www.thegreenage.co.uk/why-is-intermittency-a-problem-for-renewable-energy/

[178] finance.yahoo.com/news/winter-fury-unleashes-freeze-over-022658884.html

[179] Deely, Joe, The Energy Collective Group, energycentral.com/c/ec/can-texas-shut-down-its-remaining-coal-2030

[180] Blunt, Katherine, "The Texas freeze: Why the power grid failed", *Mint*, www.livemint.com/industry/energy/the-texas-freeze-why-the-power-grid-failed-11613800923462.html

[181] www.weather.com

[182] en.wikipedia.org/wiki/Category:Airliner_accidents_and_incidents_caused_by_ice

[183] http://www.ercot.com/gridinfo/resource

[184] See above.

[185] Angwin, Meredith; **SHORTING THE GRID: THE HIDDEN FRAGILITY OF OUR ELECTRIC GRID** (p. 212). Carnot Communications. Kindle Edition.

[186] news.energysage.com/what-is-the-power-output-of-a-solar-panel/

[187] *Prairie Climate Centre*, http://prairieclimatecentre.ca/2018/03/where-do-canadas-greenhouse-gas-emissions-come-from/

[188] www.alternative-energy-tutorials.com/energy-articles/measuring-the-power-of-a-solar-panel.html

[189] amsolar.com/diy-rv-solar-instructions/edpanelratings

[190] Thurston, Charles, "What Is The Greatest Environmental Hazard For Solar Energy? Dust!", *CleanTechnica*, cleantechnica.com/2019/07/10/what-is-the-greatest-environmental-hazard-for-solar-energy-dust/

[191] *Energy Sage*, news.energysage.com/what-is-the-power-output-of-a-solar-panel/
[192] www.currentresults.com/Weather/US/average-annual-state-sunshine.php
[193] gosolarquotes.com.au/what-time-of-day-are-solar-panels-most-efficient/
[194] EIA, www.eia.gov/todayinenergy/detail.php?id=46617
[195] Energy Education, University of Calgary, energyeducation.ca/encyclopedia/Betz_limit
[196] Atlantic County Utilities Authority, http://www.acua.com/content.aspx?id=488
[197] Merriman, Joel, "How many birds are killed by wind turbines?", *American Bird Conservatory*, abcbirds.org/blog21/wind-turbine-mortality/
[198] Eller, Donnelle, "MidAmerican Energy takes 46 wind turbines offline after blades fall in rural Iowa", *Des Moines Register*, Oct, 20, 2020, www.desmoinesregister.com/story/money/business/2020/10/20/berkshire-hathaway-midamerican-energy-windmill-blades-break-iowa-farms/5991387002/
[199] Siemens Gamesa, www.siemensgamesa.com/en-int/products-and-services/onshore/wind-turbine-sg-2-9-129
[200] www.iberdrola.com/sustainability/wind-turbines-blades
[201] Barr, Deirdra, "Modern Wind Turbines: A Lubrication Challenge", *Machinery Lubrication*, www.machinerylubrication.com/Read/395/wind-turbine-lubrication
[202] Stein, R., & Royal, T., Page 21
[203] Wilson, Robert, "Can you make a wind turbine without fossil fuels?" *Energy Central*, energycentral.com/c/ec/can-you-make-wind-turbine-without-fossil-fuels
[204] Driessen, Paul, "Life in fossil fuel free utopia", *Energy Central* energycentral.com/c/gn/life-fossil-fuel-free-utopia
[205] www.niagarafallsstatepark.com/attractions-and-tours/schoellkopf-power-plant-ruins
[206] TRVST, "The History of Hydroelectric Energy", www.trvst.world/inspiration/the-history-of-hydroelectric-energy/
[207] www.ferc.gov/sites/default/files/2020-4/HydropowerPrimer.pdf
[208] EIA, www.eia.gov/energyexplained/hydropower/tidal-power.php
[209] IEA, www.iea.org/fuels-and-technologies/bioenergy
[210] Gardner, Bruce. (2007). Fuel Ethanol Subsidies and Farm Price Support. Journal of Agricultural & Food Industrial Organization. 5. 1188-1188. 10.2202/1542-0485.1188.

211 Jacobs, Caleb, "farmers are buying up old tractors because new ones are pointlessly complicated and expensive", *The Drive*, Jan 9, 2020 www.thedrive.com/news/31761/enormous-costs-of-new-tractors-drive-demand-of-40-year-old-equipment-to-all-time-highs

212 http://energyskeptic.com/2018/power-density-of-biomass-wind-solar-requires-too-much-land-to-replace-fossil-fuels/

213 Shellenberger, Michael., **APOCALYPSE NEVER**. Harper, 2020, page 278

214 Singh, Sudheer, "One-hour rise in power outage cuts Indian households' income by 0.5%" *Energy World*, energy.economictimes.indiatimes.com/news/power/one-hour-rise-in-power-outage-cuts-indian-households-income-by-0-5-fan-zhang-world-bank/67158273

215 Angwin, Meredith. (p. 199).

216 Amelang, Soren, "Renewables cover about 100% of German power use for first time ever", *Journalism for the energy transition*, www.cleanenergywire.org/news/renewables-cover-about-100-german-power-use-first-time-ever

217 Deign, Jason, "Germany's Maxed-out Grid is Causing Trouble Across Europe", *Green Tech Media*, March 31, 2020, www.greentechmedia.com/articles/read/germanys-stressed-grid-is-causing-trouble-across-europe

218 Appunn, Kerstine, "Defining features of the Renewable Energy Act (EEG) 2014", *Journalism for energy transition*, www.cleanenergywire.org/factsheets/defining-features-renewable-energy-act-eeg

219 Wettengel, Julian, "Germany ponders how to finance renewables expansion in the future", *Journalism for energy transition*, www.cleanenergywire.org/factsheets/germany-ponders-how-finance-renewables-expansion-future

220 Amelang, Soren, "Balancing the books: Germany's "green energy account"", *Journalism for energy transition*, www.cleanenergywire.org/factsheets/balancing-books-germanys-green-energy-account

221 Worstall, Tim, *Seeking Alpha*, "SQM has a political problem – beware" https://seekingalpha.com/article/4389311-sqm-political-problem-beware

222 Timmer, John, "Comparing the actual U.S. grid to the one predicted 15 years ago", *ARS Technica*, arstechnica.com/science/2021/04/the-us-is-doing-well-on-emissions-but-not-halfway-to-zero/

223 Conley, Mike & Maloney, Tim, **ROADMAP TO NOWHERE**, 2017
224 www.nationalgeographic.com/climate-change/carbon-free-power-grid/article.html?source=carbon-free-america
225 Goldman School for Public Policy, University of California Berkley, http://www.2035report.com/transportation/wp-content/uploads/2020/05/2035Report2.0-1.pdf
226 Rossetti, Philip, "What it costs to go 100 percent renewable", *American Action Forum*, www.americanactionforum.org/research/what-it-costs-go-100-percent-renewable/#_edn1
227 https://www.ready.gov/space-weather
228 Dennis, Eric, "What the skeptics of climate catastrophe are skeptical of: Nordhaus reconsidered", *Master Resource*, www.masterresource.org/debate-issues/what-the-skeptics-are-skeptical-of/
229 Fund, John, "Professor Lockdown modeler resigns in disgrace", *National Review*, May 6, 2020 www.nationalreview.com/corner/professor-lockdown-modeler-resigns-in-disgrace/
230 Eran Bendavid and Jay Bhattacharya, "Is Covid-19 as Deadly as They Say?" *Wall Street Journal*, April 1, 2020.
231 Scott W. Atlas, "The data is in—stop the panic and end the total isolation," *The Hill*, April 22, 2020.
232 Dennis, "skeptics", *Master Resource*
233 Griggs, Mary Beth, "CERN wants to build the biggest, baddest particle collider ever", *The Verge*, Jan. 15, 2019 www.theverge.com/2019/1/15/18183828/cern-physics-particle-accelerator-hadron-collider
234 Knoploh, Sarah, "ABC Global Warming Special makes up Future", *Wall Street Journal*, www.wsj.com/articles/SB124396572915377819
235 Allon, Cap, *Electroverse*, May 5, 2021, https://electroverse.net/physicist-william-happer-there-is-no-climate-emergency/
236 Gosselin, P, *NoTrickZone*, May 17, 2013, https://notrickszone.com/2013/05/17/atmospheric-co2-concentrations-at-400-ppm-are-still-dangerously-low-for-life-on-earth/
237 Rogers, Jon, "Dark Future IMF boss warns Earth will toasted and roasted due to climate change", *The Express UK*, Oct. 25, 2017, www.express.co.uk/news/world/871180/Climate-change-IMF-Christine-Lagarde-Saudi-Arabia-inequality-dark-future

[238] *First Post*, "IMF chief warns of dark future over climate change", www.firstpost.com/world/watch-imf-chief-warns-of-dark-future-over-climate-change-4174295.html

[239] twitter.com/Survival/status/1268935324232814592

[240] Anthropocene Working Group of the International Geological Congress

[241] Chris Mooney and Brady Dennis, "The world has just over a decade to get climate change under control, UN scientists say", *Washington Post*, Oct. 7, 2018, www.washingtonpost.com/energy-environment/2018/10/08/world-has-only-years-get-climate-change-under-control-un-scientists-say/

[242] Fischetti, Mark, "We are living in a climate emergency, and we're going to say so", *Scientific American*, April 12, 2021, www.scientificamerican.com/article/we-are-living-in-a-climate-emergency-and-were-going-to-say-so/

[243] Hahn, Jason Duaine, "Scientific American Drops Climate Change to Use Climate Emergency Instead: Words Matter", *People*, April 13, 2021, people.com/human-interest/scientific-american-drops-climate-change-will-use-climate-emergency-instead/

[244] Bailey, Ronald, "Earth Day, then and now", *Reason*, May 2000, reason.com/2000/05/01/earth-day-then-and-now-2/

[245] Gwynne, Peter, "My 1975 Cooling World Story doesn't make today's climate scientists wrong", *Inside Science*, May 21, 2014 www.insidescience.org/news/my-1975-cooling-world-story-doesnt-make-todays-climate-scientists-wrong

[246] Morano, Marc, "Earth 'serially doomed': The official history of climate 'Tipping Points' began in 1864 – A new 'global warming' 12-year deadline from Rep. Ocasio-Cortez" *Climate Depot*, Jan. 22, 2019 www.climatedepot.com/2019/01/22/earth-serially-doomed-the-official-history-of-climate-tipping-points-began-in-1864-a-new-global-warming-12-year-deadline-from-rep-osasio-cortez/

[247] Marsh, George Perkins, and David Lowenthal. MAN AND NATURE. Belknap Pr. of Harvard Univ. Pr, 1864 (revised 1965).

[248] Associated Press, apnews.com/article/bd45c372caf118ec99964ea547880cd0

[249] Allon, Cap, *Electroverse*, May 7, 2021, https://electroverse.net/years-of-failed-arctic-sea-ice-predictions/

[250] Myron Ebell, Steven J. Milloy, "Wrong Again: 50 years of failed Eco-pocalyptic Predictions", Competitive Enterprise Institute, cei.org/blog/wrong-again-50-years-of-failed-eco-pocalyptic-predictions/

[251] Statista, www.statista.com/statistics/492507/concerns-about-climate-change-united-states-by-age-group/

[252] Breitbart, www.breitbart.com/politics/2019/08/24/marthas-vineyard-home-proves-obama-knows-global-warmings-hoax/

[253] Bell, Larry, "Rising tides of terror: will melting glaciers flood al gore's coastal home?" *Forbes*, June 26, 2012, www.forbes.com/sites/larrybell/2012/06/26/rising-tides-of-terror-will-melting-glaciers-flood-al-gores-coastal-home/?sh=264485aa4ee8

[254] *Conservation Conservatives*, "If Oceans are rising, Why did Gore, Obama, Kerry buy coastal mansions?" *Black & Right*, Aug. 23, 2019, www.blackandblondemedia.com/2019/08/23/if-oceans-are-rising-why-do-gore-and-obama-buy-beachfront-homes/

[255] Sheppard, Kate, "Blowing in the Wind", *Mother Jones News*, Dec. 4, 2009, www.motherjones.com/politics/2009/12/john-kerry-cape-wind/

[256] Ratcliffe, Rebecca, "Record private jet flights into Davos as leaders arrive for climate talks, *The Guardian*, Jan. 22, 2019 www.theguardian.com/global-development/2019/jan/22/record-private-jet-flights-davos-leaders-climate-talk

[257] Mead, Walter Russell, "All Aboard the Crazy Train", *Wall Street Journal*, Jan. 20, 2020, www.wsj.com/articles/all-aboard-the-crazy-train-11579554512

[258] Allon, Cap, *Electoverse*, May 7, 2021, https://electroverse.net/years-of-failed-arctic-sea-ice-predictions/

[259] McPherson, James, "The World Economic Forum: lockdowns are improving cities around the world, *Spectator Australia*, Feb. 28, 2021, www.spectator.com.au/2021/02/the-world-economic-forum-lockdowns-are-improving-cities-around-the-world/

[260] Le Quéré, C., Peters, G.P., Friedlingstein, P. et al. Fossil CO2 emissions in the post-COVID-19 era. Nat. Clim. Chang. 11, 197–199 (2021). https://doi.org/10.1038/s41558-021-01001-0

[261] International Air Transportation Association, "Deep Losses Continue into 2021", www.iata.org/en/pressroom/pr/2020-11-24-01/

[262] Jon Emont, "Countries Lose Billions Sent Home from Workers Abroad," *Wall Street Journal*, July 6, 2020.

[263] Tamny, John. WHEN POLITICIANS PANICKED: THE NEW CORONAVIRUS, EXPERT OPINION, AND A TRAGIC LAPSE OF REASON (p. 174). Post Hill Press.

[264] Abdi Latif Dahir, "135 Million Face Starvation. That Could Double," *New York Times*, April 23, 2020.

[265] Shesgreen, Deirdre, "At Earth Day climate summit, Biden promises 50% reduction in U.S. greenhouse emissions", *USA Today*, April 22, 2021

www.usatoday.com/story/news/politics/2021/04/22/president-biden-pledge-reduction-us-greenhouse-gas-emissions/7307038002/

[266] Sowell, Thomas, *The Chronicle of Higher Education*, 1996, pg. 2

[267] Landers, Peter, "Toyota's Chief Says Electric Vehicles are Overhyped", *Wall Street Journal*, Dec. 20, 2020, www.wsj.com/articles/toyotas-chief-says-electric-vehicles-are-overhyped-11608196665

[268] Shellenberger, page 32

[269] IPCC, "Making Peace with Nature", 2021 page 26.

[270] Reuter, Dominick, April 30, 2021, *Business Insider*, https://www.businessinsider.com/electric-car-owners-switching-gas-charging-a-hassle-study-2021-4

[271] Woollacott, Emma, *BBC News*, April 27, 2021, https://www.bbc.com/news/business-56574779

[272] Blain, Loz, "Oceanbird's huge 80-meter sails reduce cargo shipping emissions by 90%", *New Atlas*, Sept. 14, 2020, newatlas.com/marine/oceanbird-wallenius-wing-sail-cargo-ship/

[273] The Maritime Executive, "Unique Sail Cargo ship departs on first Atlantic crossing from France", Nov. 20, 2020 www.maritime-executive.com/article/unique-sail-cargo-ships-departs-on-first-atlantic-crossing-from-france

[274] New Altas, *Oceanbird*

[275] Shellenberger, page 188

[276] Carrington, Damian, "World's largest all-electric aircraft set for first flight", *The Guardian*, May 27, 2020, www.theguardian.com/world/2020/may/27/worlds-largest-all-electric-aircraft-set-for-first-flight

[277] Ekins, Emily, "68% of Americans wouldn't pay $10 a month in higher electric bills to combat climate change", *Cato Institute*, March 8, 2019 www.cato.org/blog/68-americans-wouldnt-pay-10-month-higher-electric-bills-combat-climate-change

[278] ibid

[279] Stein, Ronald, "NIMBYS are making more noise than wind turbines", *New Geography*, Dec. 15, 2020 http://www.newgeography.com/content/006880-nimbys-are-making-more-noise-than-wind-turbines

[280] Ball, Jeffery, "Renewable Energy, Meet the new Nimbys", *Wall Street Journal*, Sept. 4, 2009, www.wsj.com/articles/SB125201834987684787

[281] Emerson, Sandra, "It's lights out on big solar in San Bernardino county desert", *The Sun*, March 4, 2019, www.sbsun.com/2019/02/28/san-

bernardino-county-board-to-prohibit-renewable-energy-development-in-key-desert-areas/

[282] Shellenberger, page 199

[283] Huston, Warner Todd, "Leftist, Antifa woman charged with terror plot to derail train in Washington state", *Great American Politics*, Dec. 3, 2020, greatamericanpolitics.com/2020/12/leftist-antifa-woman-charged-with-terror-plot-to-derail-train-in-washington-state/

[284] Adl-Tabatabai, Sean, "Police: antifa terrorists derailed amtrack train in Washington" *New Punch*, Dec. 19, 2017, newspunch.com/antifa-derailed-amtrak-train-washington/

[285] *HeadlineUSA*, May 9, 2021, https://headlineusa.com/cyber-attack-halts-major-u-s-pipeline-system-threatens-gas-prices/

[286] Service, Robert F. "Can the world make the chemicals it needs without oil?", *Science*, Sept. 19, 2019, www.sciencemag.org/news/2019/09/can-world-make-chemicals-it-needs-without-oil

[287] Rossetti, "what it costs to go 100 percent renewable"

[288] *Inspire*, www.inspirecleanenergy.com/blog/clean-energy-101/cost-of-renewable-energy

[289] *Twist Bioscience*, www.twistbioscience.com/blog/perspectives/those-pesky-petrochemicals-there-better-way

[290] Ferry, David, "The Promises and Perils of Synthetic Biology", *Newsweek*, March 11, 2015, www.newsweek.com/2015/03/20/promises-and-perils-synthetic-biology-312849.html

[291] Sowell, Thomas, **THE QUEST FOR COSMIC JUSTICE**, Touchstone, New York, NY, 1999, page 86

[292] en.wikipedia.org/wiki/White's_law

[293] www.un.org/development/desa/dpad/least-developed-country-category.html

[294] Stein, R. Royal, T., Page 38

[295] Wang, Brian, "Developed country Population from 17% to 50% of world by 2050", *nextBIG Future*, Nov 19, 2018, www.nextbigfuture.com/2018/11/developed-country-population-from-17-to-over-50-of-world-by-2050.html

[296] Shellenberger, page 229

[297] Stein, R. Royal, T., Page 38

[298] Chamary JV, "The Science of Avengers:Endgame proves Thanos Did Nothing Wrong", *Forbes*, May 7, 2019, www.forbes.com/sites/jvchamary/2019/05/07/avengers-endgame-biodiversity/?sh=561998fc775b

[299] Qu G. "Population control and environment protection". Renkou Yanjiu. 1982 Jan 29;(1):43-8. Chinese. PMID: 12338285

[300] *Center for Biological Diversity,* www.biologicaldiversity.org/programs/population_and_sustainability/population/

[301] en.wikipedia.org/wiki/2010_eruptions_of_Eyjafjallaj%C3%B6kull

[302] Goyal, Radhika, "Iceland's Most Active Volcano is Likely Headed for Another Eruption", Columbia Climate School, *State of the Planet*, Aug. 4, 2020, blogs.ei.columbia.edu/2020/08/04/iceland-volcano-nearing-eruption/

[303] United States Geological Survey, www.usgs.gov/faqs/how-much-ash-was-there-may-18-1980-eruption-mount-st-helens?qt-news_science_products=0#qt-news_science_products

[304] University Corporation for Atmospheric Research, "Mount Tambora and the Year without a Summer", scied.ucar.edu/learning-zone/how-climate-works/mount-tambora-and-year-without-summer

[305] www.dlr.de/tt/fluginsekten

[306] Davies, Samantha, "Saharan Dust over North Texas this Weekend", *NBCDFW*, June 19, 2020, www.nbcdfw.com/weather/weather-connection/saharan-dust-could-bring-spectacular-sunsets-to-texas-gulf-coast-next-week/2391700/

[307] Karlsruhe Institute of Technology. "Saharan dust: Reliable forecasts for photovoltaic output." *ScienceDaily*. 27 July 2016. <www.sciencedaily.com/releases/2016/07/160727090604.htm>

[308] Villagomez, Phil, "Wildfire Ash Impacts on Solar PV Systems", *TerraVerde Energy*, Oct. 01, 2020, terraverde.energy/wildfire-ash-impacts-on-solar-pv-systems/

[309] *Sonoma Clean Power*, "Local School Districts to Receive Free Battery Storage Assessments", sonomacleanpower.org/news/local-school-districts-can-receive-free-battery-storage-assessments

[310] Thornton, John P, "The Effect of Sandstorms on PV arrays and components," 1992

[311] Noon, Chris, "Just Deserts: Oman's New Wind Turbine Can Handle Sandstorms and Desert Sun", General Electric, Dec. 28, 2018, www.ge.com/news/reports/just-deserts-wind-turbine-can-handle-sandstorms-desert-sun

[312] Stein, R., & Royal, T., Page 64

[313] Strachan H. **THE FIRST WORLD WAR**. Simon and Schuster New York City, NY (2014)

[314] watson.brown.edu/costsofwar/papers/ClimateChangeandCostofWar Crawford, Neta, Pentagon Fuel Use, Climate Change and the Cost of War Final.pdf June 12, 2019

[315] www.globalsecurity.org/military/systems/ground/m1-specs.htm
[316] Schmidt, Bridie, "An Electric Tank? Army see multiple advantages in dumping fossil fuels", *The Driven*, thedriven.io/2019/09/10/an-electric-tank-army-sees-multiple-advantages-in-dumping-fossil-fuels/
[317] www.cliftonsteel.com/military-steel
[318] www.cmc.com/en-us/what-we-do/america/performance-steel/trushield
[319] www.ssab.com/products/brands/armox
[320] www.leecosteel.com/steel-plate-products/military-armor-ballistic-steel-plate/
[321] https://climateandsecurity.files.wordpress.com/2019/implications-of-climate-change-for-us-army_army-war-college_2019.pdf
[322] www.ipaa.org/independent-producers/
[323] www.api.org/media/files/oil-and-natural-gas/tankerstankers-lores.pdf
[324] www.eia.gov/tools/faqs/faq.php?id=29&t=6
[325] www.americangeosciences.org/geoscience-currents/transportation-oil-gas-and-refined-products
[326] www.fueleconomy.gov/feg/quizzes/answerquiz16.shtml
[327] Kingson, Jennifer A. "Cities are starting to ban new gas stations", *Axios*, Mar. 1, 2021, www.axios.com/cities-ban-gas-pollution-fb61cf2f-9893-466f-9559-9c4d27bbc3b6.html
[328] Lovins, A.. "Energy Strategy: The Road Not Taken?" *Foreign Affairs* 55 (1976): 65.
[329] *BBC*, "Denmark to build first energy island in North Sea", Feb. 4, 2021, www.bbc.com/news/world-europe-55931873
[330] Investor's Business Daily, Feb. 10, 2015, https://www.investors.com/politics/editorials/climate-change-scare-tool-to-destroy-capitalism/
[331] Epstein, A, page 36
[332] Stein, R., & Royal, T., Page 69
[333] Shellenberger, page 4
[334] Heller, Tony, "NOAA Data Tampering Approaching 2.5 degrees", *Real Climate Science*, March 20, 2018, realclimatescience.com/2018/03/noaa-data-tampering-approaching-2-5-degrees/
[335] Hinderaker, John, *Powerline*, June 2, 2021 https://www.powerlineblog.com/archives/2021/06/where-has-all-the-global-warming-gone.php
[336] Joan Martinez-Alier, Ronald Muradian, **HANDBOOK OF ECOLOGICAL ECONOMICS,** 2015, Page 130
[337] Bailey, R. **THE END OF DOOM,** 2015

[338] Boissoneault, Lorraine, "The Cuyahoga River Caught Fire at Least a Dozen Times, but No One Cared Until 1969", *Smithsonian*, June 19, 2019, www.smithsonianmag.com/history/cuyahoga-river-caught-fire-least-dozen-times-no-one-cared-until-1969-180972444/

[339] Jacopo Pasotti and Elisabetta Zavoli, "People Are Living Inside Landfills As The World Drowns In Its Own Trash", *Huffington Post*, Oct. 23, 2018

[340] blog.smu.edu/smumagazine/2018/02/bringing-fresh-water-bolivian-village/

[341] www.cdc.gov/healthywater/global/wash_statistics.html

[342] ourworldindata.org/energy-access

[343] cities4forests.com/lg-se-social-equity/

[344] Samuel Asumadu Sarkodie, Vladimir Strezov, "Empirical study of the Environmental Kuznets curve and Environmental Sustainability curve hypothesis for Australia, China, Ghana and USA", *Journal of Cleaner Production*, Volume 201, 2018, Pages 98- 110, www.sciencedirect.com/science/article/abs/pii/S0959652618323722

[345] Billen, G., Barles, S., Chatzimpiros, P. et al. *Grain, meat and vegetables to feed Paris: where did and do they come from? Localising Paris food supply areas from the eighteenth to the twenty-first century.* Reg Environ Change 12, 325–335 (2012). doi.org/10.1007/s10113-011-0244-7

[346] Sayet, E, **THE WOKE SUPREMACY**, 2020 page 139.

[347] EPA, www.epa.gov/air-trends

[348] https://granitegrok.com/blog/2021/06/it-isnt-the-wind-power-they-want-its-the-money-subsidies-in-their-pockets

[349] Bailey, R., Page 232

[350] Easterbrook, G., page 21

[351] Bailey, R., Page 216

[352] Nordhaus, Willam, "An Analysis of the Dismal Theorem", Cowles Foundation for Research in Economics, Yale University

[353] See above.

[354] Easterbrook, G. page 22

[355] Vermeulen, Freek, "Many Strategies Fail Because They're Not Actually Strategies", *Harvard Business Review*, Nov. 08, 2017

[356] https://www.zerohedge.com/energy/dutch-court-orders-shell-aggressively-cut-carbon-emissions-landmark-decision

Index

Accord, Paris Climate, 37, 41, 201, 231
Africa, 44- 45, 158, 224, 235
Agency, Environmental Protection, 83, 104, 142, 254
Aircraft, 70-72, 208
  Boeing 747, 217
  Cessna, 218
  Concorde, 60
  Solar Impulse, 217
Alaska, 85, 178
Amazon, 42
American Petroleum Institute, 3, 83
Angwin, Meredith, 136, 141
Army War College, 15-16, 241
Asia, 99, 217, 235
Asphalt, 78
Aspirin, 119
Associated Press, 220
Association
  American Gas, 33
  American Psychological, 7, 11
Atlantic County Utilities Authority, 149
Attenborough, Sir David, 44, 192, 231
Bacon, Francis, 46
Bailey, Ronald, 51, 133, 262
Band, Coldplay, 42
Bangladesh, 101, 134, 206, 231

Bayer, 90, 109, 112
Baylor Medical Center, 99
Bias, Confirmation, 11-12
Bias, Status Quo, 11-12, 44, 252
Bicycle, 118
Biden Unity Plan, 24
Biden, President Joseph, 6, 23, 30, 32, 37-39, 45, 71, 166, 174, 182, 193, 201, 202, 207-208, 242, 248, 266
Birkenstock, 89
Blackrock Investments, 38
Bolin
  Benjamin, 47
  Samuel, 47
  William, 47
Bono, 42
Books
  A Question of Power Electricity and the Wealth of Nations, 136
  Circles, 90
  Dark Age America, 7, 22
  Global Warming, 20
  How to Avoid a Climate Disaster, 266
  Just Green Electricity: Helping Citizens Understand a World without Fossil Fuels, 3, 50, 88, 108
  Limit to Growth, 188, 254
  Man and Nature, 195
  Population Bomb, 34, 111

Shorting the Grid, 136, 141
The End of Doom, 262
The Long Winter, 54
The Nitpickers Guide to Star Trek, 120
The Scout Mindset, 16
The Woke Supremacy, 120
When Politicians Panicked, 8, 137
Brazil, 155, 253, 256
Break Free Movement, 88
Breakthrough Energy, 41
Brooklyn Bridge, 93
Brown, Robert G., 137
Bryce, Robert, 136
Burke, James, 90
Butylene glycol, 114
Cairo, 253
California, 25, 144, 171, 178, 201, 216, 235, 245
   San Bernardino, 222
   San Francisco, 62, 217, 218, 228
   Santa Monica, 62
   Sonoma Valley, 236
   Stockholm, 62
Cameroon, 158, 230, 231
Cape Wind, 201
Carpet, 96
Chernobyl, 31, 61
China, 57, 91-92, 131, 137, 145, 155, 171, 212
Citibank, 38
Civilian Conservation Corp, 153
Clausewitz, Carl von, 237
CLEAN Future Act, 24, 39
Clifton Steel, 239
Clinton, Hillary, 186
CMC, 239
Coalition Opposing New Gas Stations, 243
Colorado Oil and Gas Association, 106
Colorado, Aspen, 62
Coltura, 243
Competitive Enterprise Institute, 197
Conservation International, 36
Conference of the Parties (COP), 24
Copper, 151, 180, 198
Coronavirus, 91, 131, 186-187, 204, 206, 262
Corporate Average Fuel Economy, 142
Crude Oil
   Brent, 82
   MARS, 82
   OPEC, 82
   WTI, 82
Current
   Alternating, 28-30
   Direct, 28-29, 32, 142, 199, 217-218, 229
Cuyahoga R., 253
D'Souza, Dinesh, 132
Dacron, 102, 107
Daimler, Gottlieb, 87
Dam, Grand Coulee, 25
Dam, Hoover, 25, 153
Deere, John, 47
Delaware
   Rehoboth Beach, 202
Denmark, 165, 244

Density, Energy, 154, 155
Dino, 74
Dinosaurs, 74-76
Dismal Theorem, 262
Dow Chemical, 117
Drake, Col. Edwin, 77, 86
Easterbrook, Gregg, 263
Ebell, Myron, 197
EcoWatch, 62
Edison, Thomas, 28, 60
Ehrlich, Paul, 111
Electricity Reliability Council of Texas (ERCOT), 138, 140, 141, 142, 162, 243
Emanuel, Kerry, 247
Energy Information Administration, 147, 169
Epoxy, 118
Epstein, Alex, 111, 246
Ethanol, 59, 67, 71, 155, 183
Ethiopia, 253
Everest, Mt., 105
Eyjafjallajokull, Mt., 234, 235
Fagradalsfjall, Mt., 234
Fair, DeLome, 91
Farman, Joseph, 263
Farrand, Phil, 120
Federal Energy Regulatory Commission, 153
Feedstocks, 70, 72, 109, 111, 113, 183-184, 212, 227
Ferguson, Neil, 186-187
Fink, Larry, 38
Flatiron Building, 93
Florida, 71, 186, 211
Florida International University, 186
France, 153, 256

Francis, Pope, 193
Fuel
  Aviation, 87, 266
  Diesel, 59, 60, 87, 118, 155
  Kerosene, 59, 71, 86
  Primary, 56-59
  Propane, 59, 68, 81, 155, 271
  Secondary, 59-61, 70-72
Fukushima, 31, 61
Future Shock, 2
Galef, Julie, 6, 16
Gates, Bill, 37, 41, 42, 94, 193, 266
Geothermal, 25
Germany, 31, 165, 211, 212, 228, 237
Gore, Albert Jr, 199, 201, 248
Gore-Tex, 106, 107
Grayson, Rep. Alan, 27
Green New Deal, 24, 26, 27, 28, 29, 33, 166
Green Solution, 24, 26, 28, 29, 37, 44, 45, 46, 50, 64, 71, 72, 137, 157, 159, 163, 206, 217, 219, 246, 248
Greenland, 8
Greenpeace, 24, 32
Greer, John Michael, 7, 22
Hamlet, 101
Hillary, Edmund, 105
Hoffman, Felix, 109
Hybrid Vehicle, 73
Hydroelectric, 25, 152, 164
Iceland, 234
IG Farben, 109
Imperial College, 186, 187

Independent Petroleum Association of America, 39
India, 42, 91, 109, 131, 132, 145, 211, 253
International Energy Agency, 37, 168
International Monetary Fund (IMF), 192
IPCC, 35-36, 44, 131-133, 193, 196, 212, 230, 267
Japanese Meteorological Agency, 15
Jefferson, Martha, 104
Jersey-Atlantic Wind Farm, 149
Just Facts, 21
Kahan, Dan, 13
Kampala, 253
Kennedy, Robert F. Jr, 163
Kennedy, Senator Ted, 201
Kerry, John, 38, 193, 201, 221
Kier, Samuel, 76, 108
KlimaFa, 42
Kuznets Curve, 254
Lagarde, Christine, 192
Least Developed Countries (LDC), 229
Leeco Steel, 239
Liberty Mutual, 46
Life Expectancy, 122
Liquid Petroleum Gas (LPG), 71, 269
Llojila Grande, 254
London, 186, 217, 254, 263
Lovins, Amory, 244
Lubricants, 70, 72
M1A1 Abrams, 239
Macintosh, Charles, 91

Mad Cow Disease, 186

Magazine
  Business Week, 222
  Discover, 20
  Nature – Climate Change, 204
  The Atlantic, 2, 197
Mahbubani, Kishore, 20
Malaysia, 253
Marchetti, Cesare, 52
Marsh, George Perkins, 195
Massachusetts,
  Bridgewater, 18
Maybach, William, 87
McLaren, Duncan, 43
Merrill Lynch, 38
Milloy, Steven J, 197
Mississippi, 153
Mortgages, 38
Myanmar, 253
Nairobi, 253
Naphtha, 91
Napoleon, 21- 22
NASA, 15, 250
National Climate Bank, 39
National Geographic Society, 36, 102, 234
National Renewable Energy Laboratory, 143, 173, 236
Natural Disasters, 133
Natural Resource Defense Council, 36
Neoprene, 107
Netherlands, 165
Neumann, Caspar, 75
New York

New York City, 93, 189, 199
Niagara Falls, 152
New York Times, 21, 197
NIMBY, 221-222, 227
Nixon, President Richard, 248
Nordhaus, William, 262
Norgay, Tenzing, 105
North Face, 106
Nye, Bill, 193
Nylon, 107, 118
Obama, President Barak, 193, 201, 207
Ocasio-Cortez, Rep. Alexandria, 193, 248
Olympics, 199, 202
Oman, 237
Opus12, 226
Otto, Nicolaus, 87
Owen, Sir Richard, 76
PCAF, 38, 40
Pennslyvania
   Pittsburgh, 76, 93
Philancapitalism, 42
Plague, Black Death, 263
Poland, 165
Polyethylene Glycol, 115
Polyvinyl Chloride, 96
Poverty Line, 130
Propylene Glycol, 114-116
Prudhoe Bay, 85
Rabbit Clothing, 104
Rail
   Amtrak, 223
   Burlington Northern Santa Fe, 223
   Eurostar, 217
Reactors, nuclear, 31, 269

Reagan, President Ronald, 265
River, Raccoon, 48
Rocket, 154
Romania, Ploiesti, 238
Roosevelt, President Franklin, 60
Royal, Todd, 3, 6, 50, 88, 108, 121, 197, 230
Rubber, 89, 91, 95, 97-98, 106-107, 118-119
Rugs, 118
Sahara, Wind Storms, 158, 235, 237
Samoa Air, 98
Sanders, Sen. Bernie, 193
Sayet, Evan, 120, 257
Schneider, Stephen H, 20
Schoellkopf Power Station, 152
Scientific American, 193
Service, Robert F., 92
Shellenberger, Michael, 157, 268
Ship
   Cruise Liners, 30
   Grain de Sail, 215
   Mayflower, 58, 60
   Merchant, 30, 214
   Neoline, 215, 216
   Oceanbird, 215, 216
   SS Great Western, 58
   SS United States, 58
Siemens, 146-147, 150-151
Sinclair Oil Company, 74-75
Smil, Vaclav, 157
Solutions Project, 24, 27, 32, 63, 166, 171

Sonoma Clean Power, 236
South Korea, 153, 269
Southern Methodist University, 253
Sowell, Thomas, 207, 228
SpaceX, 107
Sri Lanka, 101
SSAB, 239
St. Helens, Mt., 234-235
Stand.earth, 243
Standard Test Conditions(STC), 143
Starbucks, 222
Stein, Ronald, 3, 6, 50, 88, 108, 121, 230
Steyer, Tom, 193
Stott, Philip, 195
Strachan, Sir Hew, 238
Switzerland
  Davos, 202
Taiwan, 100
Taiwan Semiconductor Manufacturing Company, 100
Tambora, Mt., 235
Tamny, John, 8, 137
Teflon, 109
Tennessee Valley Authority, 25
Texas, 25, 27, 46, 63, 77, 82, 138-141, 148, 162, 211, 223, 243
Thunberg, Greta, 193
Toffler, Alvin, 2
Trump, President Donald, 23
Trump, Melania, 101
Tzu, Sun, 219

UNCCD, 236
United Nations, 4, 24, 43, 54-55, 131, 133, 192, 196-199, 229, 236, 253, 267
  Environmental Program, 196
Uranium, 59-61
US Tire Manufacturer Association, 97
USDA, 37
Vaseline, 114
Vatican, 42
Vermeulen, Freek, 263
VF Corporation, 106
Vietnam, 101
Virgin Atlantic, 98
Volatile Organic Compounds (VOC), 95
Volcano, 234, 236
Wall Street Journal, 137
Washington, 32, 71, 144, 197, 217, 218, 223, 234, 265
Washington, President George, 265
Weitzman, Martin, 262
Wilder, Laura Ingalls, 54
World Bank, 130-131, 159
World Economic Forum (WEF), 204-205, 211, 244
World Health Organization, 55, 111
World War I, 110, 237, 248
World War II, 110, 172, 173, 238, 248
Yale, Cultural Cognition Project, 13

www.ingramcontent.com/pod-product-compliance
Lightning Source LLC
Chambersburg PA
CBHW071446220526
45472CB00003B/683